D0301847

Cost Planning of PFI and PPP Building Projects

Also available from Taylor & Francis

Public Private Partnerships in Construction
D. Cartlidge

Procurement in the Construction Industry
W. Hughes et al.

Financing Construction
R. Kenley

Project Planning and Control
D. Carmichael

Construction Project Management
P. Fewings

Risk Management in Projects 2nd ed.
M. Loosemore et al.

Construction Collaboration Technologies
P. Wilkinson

Cost Planning of PFI and PPP Building Projects

Abdelhalim Boussabaine

NEW YORK AND LONDON

First published 2007
by Taylor & Francis
2 Park Square, Milton Park, Abingdon, Oxon OX14 4RN

Simultaneously published in the USA and Canada
by Taylor & Francis
270 Madison Ave, New York, NY 10016

Taylor & Francis is an imprint of the Taylor & Francis Group, an informa business

Typeset in Sabon by
RefineCatch Limited, Bungay, Suffolk
Printed and bound in Great Britain by
TJI Digital, Padstow, Cornwall

British Library Cataloguing in Publication Data
A catalogue record for this book is available
from the British Library

Library of Congress Cataloging in Publication Data
A catalog record for this book has been requested

ISBN10: 0–415–36622–4 (hbk)
ISBN13: 978–0–415–36622–9 (hbk)

Read! In the Name of your Lord Who has created (all that exists). He has created man from a clot. Read! And your Lord is the Most Bountiful, Who has taught (writing) by the pen. He has taught man that which he knew not.

Qur'an

Say: 'O my Lord! Increase me in knowledge.'

Qur'an

Contents

Figures

Tables

Preface

In recent decades, the construction industry has witnessed unprecedented growth in the use of the Private Finance Initiative (PFI) procurement route in UK projects such as schools, hospitals, prisons and civil infrastructure. As a result, a need for metrics and methods aimed at improving the cost and performance management of PFI and public–private partnerships (PPP) projects has arisen. The availability of these techniques is an essential part of the overall management of PFI/PPP projects, while additionally enabling stakeholders to contribute effectively to UK sustainable policies within the built environment.

In contrast to conventionally financed project procurement, the PFI/PPP approach has challenged the way stakeholders prepare and arrange budgets. Presently, there has been little available on how PFI/PPP projects are costed. In this book, cost planning is central to both the understanding and analysis of PFI/PPP schemes. Hence, this book sets out to explain how PFI/PPP cost appraisal issues can be appreciated by means of the correct application of innovative costing methods, where the emphasis is on planning and control. The aim of this book is to introduce cost planning of PFI/PPP building projects to both professionals and students in the construction industry and beyond. This is a tool that will provide students and practitioners with a structural and conceptual basis for cost planning of PFI/PPP schemes.

The book has 13 chapters divided into five Parts. Each section is not presented in isolation and an integrated approach has been adopted. Part I consists of three chapters and provides an introduction to the rationale of the rise of PFI/PPP schemes in the UK construction economy. Chapters 2 and 3 call for changes in the way that the whole life cycle value is perceived, created and exchanged throughout the life cycle of building assets. If the theories advocated by these chapters are taken forward by the built environment stakeholders, this could have a significant impact on the way value for money analysis is carried out. Part II has three chapters and provides an introduction to procuring building assets under the PFI/PPP model as well as focusing on the foundation of the theoretical understanding of the organisational structure of different project funding mechanisms and time value-based investment concepts. Part III consists of three chapters and is aimed at explaining the process of cost planning in

PFI/PPP projects. Cost planning from the viewpoint of purchasers and providers is presented and explained with examples in Chapters 8 and 9 respectively. The process is radically different from the traditional project investment appraisal. Traditional cost planning processes are linear and exclude a large proportion of whole life cycle budget planning aspects that are essential in PFI/PPP schemes. The challenge for the cost analyst is to develop whole life cycle cost plans with a minimum of detailed design and information. Cost plans are negotiated at the pre-financial close stage. At the financial close stage, all prices, funding terms and payment mechanisms are confirmed and fixed. Part IV consists of three chapters and addresses cost variation at both pre- and post-financial close stages. The author's emphasis here is on the process of setting cost control benchmarks and the significance of governance in monitoring and responding to cost variations. Finally, in Part V the book focuses on risk and uncertainty in costing PFI/PPP schemes. I believe that the pricing of risk is a progressive and iterative process that needs to be considered cautiously by risk analysts.

This text endeavours to provide the reader with existing knowledge as well as present innovative thinking for future development and management of PFI/PPP cost planning processes. Hence, readers may find some concepts, ideas and procedures that are not common in professional practice. It is vital that all built environment stakeholders should be enthusiastic in order to be innovative and creative in the way we deliver value to our clients and to ourselves. Although the book has tried to cover as much as possible regarding aspects of cost planning of PFI/PPP projects, guidance on suitable sources of additional information is provided. Readers who are eager to explore issues presented in this book in greater depth should consult the list of further reading and the Bibliography at the end of the book.

This book is presented to readers based on the belief that no matter how much knowledge you acquire, there will always be someone who knows more than you.

Abdelhalim Boussabaine
Liverpool University
July 2006

Acknowledgements

Tables 8.2, 8.3, 8.4 and 8.5 are reproduced from National Health Service Estates (NHS) (1995) *Capital Investment Manual: Management of Construction Projects*, under the terms of Crown Copyright Policy Guidance issued by HMSO.

I acknowledge and thank J. Lewis and Dr R. Kirkham for their assistance in delivering this book and their helpful comments on Chapters 2 and 3. I would also like to express my sincere appreciation to Ian Hunter and Chris Belledonne of the Liverpool School of Architecture, who assisted me continually throughout the preparation of this book. My thanks to Katy Low, assistant editor at Taylor & Francis, for her support. I thank Dr Dean, Dr Sally and my wife for proofreading the first draft of this book. I am grateful to 4Ps and J. Lewis for their permission to reproduce Table 13.1 and Figure 3.7.

Researching and writing this book was exceedingly time-consuming. I am indebted to my wife, Ula, my daughter, Emaan, and my sons, Mohammed and Youcef, for their patience, support, encouragement, and good humour during this period.

During the course of researching this book I have reviewed hundreds of articles and documents. I would like to apologise if any of them were inadvertently not cited or acknowledged. I also affirm that any mistakes and errors in the book are entirely my responsibility.

Abbreviations

4PS	Public–Private Partnerships Programme
ADSCR	annual debt service cover ratio
AF	annuity factor
BAFO	best and final offer
BCIS	building cost information service
BLT	build/lease and transfer
BOOT	build, own, operate and transfer
BOT	build, own and transfer
BRT	build/rent and transfer
CGF	credit guarantee finance
CHP	central heating plant
CSS	central sterile service
DBFO	design, build, finance and operate
DBM	design, build and maintain
DCAG	Departmental Cost Allowance Guide
DCF	discounted cash flow
DSCR	debt service cover ratio
EAC	equivalent annual cost
EBITDA	earnings before lease charges, interest, taxes, depreciation and amortisation
EPC	engineer procure construct
ERIC	Estates Returns Information Collection System
EVA	economic value added
FBC	full business case
FC	financial close
FCP	final commissioning programme
FITN	final invitation to tender notice
FM	financial management
FRS	financial reporting standard
FTN	final tender notice
GDB	gross domestic balance

GFA ground floor area
HMT Her Majesty's Treasury
HV high voltage
ICE Institution of Civil Engineers contract
IRR internal rate of return
IT independent tester
ITN instructions to negotiate
JCT joint contracts tribunal
KPI key performance indicator
LCA life cycle assessment
LLCR loan life cover ratio
LV low voltage
MES mechanical and electrical service
MPIS Median Index of Public Sector Building Tender Prices
MVA market value added
NAO National Audit Office
NEC new engineering contract
NPC net present cost
NPV net present value
OBC outline business case
OFA occupied floor area
OGC Office of Government Commerce
OJEC *Official Journal of the European Community*
OJEU *Official Journal of the European Union*
PA planning approval
PB preferred bidder
PBM payback method
PFI private finance initiative
PITN preliminary invitation to negotiate
PLCR project life cover ratio
PPP public–private partnership
PQQ pre-qualification questionnaire
PSC public sector comparator
PUK Partnerships UK
PV present value
RPI retail price index
SOC service output specifications
SoPC Standardisation of PFI Contracts
SPV special purpose vehicle
TSR total shareholder return
VAT value added tax
VFM value for money
VM value management

WACCS weighted average cost of capital
WLC whole life cycle
WLCC whole life cycle costs

Part I

PFI/PPP and VFM rationale

Chapter 1

The rationale behind the private financing of public projects

Introduction

The UK Government introduced the Private Finance Initiative (PFI) as a policy to allow and regulate privately financed public projects. The private sector playing a part in the financing, creation and operation of public services and built assets is not a new idea, as this initiative can be traced back over several decades. This chapter outlines the rise of PFI/PPP (public–private partnerships) in the UK construction economy. It puts into perspective the rationale for and against the PFI procurement method and sets out some basic definitions and clarifies the difference between PPP and PFI terminology. It points out the benefits of public–private partnerships, it discusses the value of cost planning PFI/PPP projects and explains the challenges that are facing cost analysis in pricing long-term contracts. The chapter hopes to lay the ground for the subsequent chapters where the real in-depth analysis of the issues surrounding the cost planning of PFI/PPP projects lies.

What is PFI/PPP?

Lack of uniform agreement on the meaning of PFI/PPP creates confusion, introduces ambiguity and may lead to contractual disputes. If a precise definition is not agreed, it will be very difficult to measure the success or otherwise of this mode of project procurement and financing. Hence, this section presents the definition of PFI and PPP. Public–private partnerships (PPP) were introduced by the UK Government to deliver modern, high-quality public services and to promote the UK's competitiveness. Any collaboration between the public authorities and the private sector is referred to as a partnership. Partnering is defined by the National Audit Office (NAO 1999a) as: 'The situation where a public organisation and a private one work together to provide a service with some sharing of risk and reward, usually over a period of time.'

According to the Treasury (HMT 2000a) PPPs are a form of long-term relationship between the public and private sector for the benefit of both parties and cover a wide range of business partnership arrangements including joint ventures, concessions, sale of equity stakes in state-owned businesses, PFIs, franchises and outsourcing of public services. 4Ps (2006) defined and explained the purpose of PPP as follows:

> Public–private partnerships (PPPs) are a generic term for the relationships formed between the private sector and public bodies often with the aim of introducing private sector resources and/or expertise in order to help provide and deliver public sector assets and services. The term PPP is used to describe a wide variety of working arrangements from loose, informal and strategic partnerships to design, build, finance and operate (DBFO) type service contracts and formal joint venture companies.

It is clear from the above that PPP is broader in concept than PFI. But PFI is considered as a form of PPP. The House of Commons Select Committee on Treasury (2002) defined PFI as follows:

> The principle of PFI is that a public sector body (central or local government or part of the health service, generically called 'the authority') obtains a service rather than an asset. A private sector contractor (often specially created for the purpose) funds any asset required and is then paid for the service provided. Usually payments will be made by the commissioning authority, but in some projects (e.g. a toll bridge) they are made by the public. There is also the possibility of charges to the public being subsidised by payments by the commissioning authority (e.g. a railway where the fares are subsidised for public benefit reasons). Normally, therefore, the commissioning authority will avoid the need for capital expenditure at the beginning of the project in exchange for making payments for the service as it is delivered, often over a period of up to thirty years. The private finance is temporary: the public sector still pays in the end. Because the contractor is paid for the service provided, it should be possible to arrange that he is taking the risks inherent in the construction of the asset.

In 1995, the UK Chancellor of the Exchequer gave the following description of PFI in his Budget speech:

> Under the PFI the public sector does not simply sign a contract to buy a prison, a train or a computer system. It pays to have specific services supplied at guaranteed levels of performance. The government chooses the quality services the public require, and then goes out and acquires those services from private companies with the finance and expertise to deliver.

The Private Finance Panel (1995) defined PFI as follows:

> PFI has become one of the government's main instruments for the delivery

> of high quality and cost effective public services. It enables value for money and improvements to be obtained by requiring the private sector in competition to be innovative in the design and operation of asset-based services, manage an appropriate level of risk and adequately maintain assets on a long-term basis.

The NAO definition emphasises the issues of procurement, management of public services and lack of funding for public projects:

> PFI is now established as a major form of Government procurement . . . PFI has introduced changes in the way public services are procured and managed.
>
> (NAO 2001)

> A policy introduced by the Government in 1992 to harness private sector management and expertise in the delivery of public services . . .
>
> The PFI approach can enable departments to undertake projects which they would be unable to finance conventionally.
>
> (NAO 1999a)

Broadbent and Laughlim (2003) explained:

> PFI, in its purest form, is a design build finance and operate (DBFO) system. It usually involves the provision, by a private sector consortium, of property-based services for a period of a minimum 30, and, more usually, 60 years, to a public sector 'purchaser'. In exchange for these services over this 30/60-year time horizon, the public sector pays a monthly, in effect, lease cost to the private sector supplier. This monthly cost is revised periodically as the contract progresses.

It is apparent from the above quotations that PFI evolved from the development of public–private partnerships. Hence, hereafter, in this book, PFI and PPP are treated synonymously.

In examining the above quotations on PFI and PPP, one can identify the following key concepts that lead to the emergence of PFI/PPP:

- sharing risk and reward;
- introduction of the resources and expertise of the private sector into the public sector;
- delivery of public services by the private sector at a guaranteed level of performance;
- delivery of high quality and cost-effective public services;
- innovation in the design and operation of assets;
- risk transfer;

- introduction of a new form of managing and procuring public services and projects;
- long-term contractual agreements between private and public sectors;
- a substitute for public expenditure control;
- value for money in the delivery of public services.

The rationale for PFI/PPP has evolved over time to address these issues. While some of these concepts were instrumental in the materialisation of PFI/PPP, more recently the emphasis has shifted towards the theory that PFI/PPP is a public procurement system that can deliver value for money and manage risk transfer to the benefit of the public sector. Hence, this chapter seeks to provide an overview of PFI/PPP and the arguments relating to it.

The rise of PFI/PPP in the UK construction economy

The rise of PFI/PPP in the Western construction economy is not a new phenomenon. Many Western countries have used this procurement method to build motorways, prisons and hospitals. For example, in France, private financing for motorways was introduced in the late 1960s and early 1970s. Similar initiatives were also undertaken in Spain and Italy in the early 1970s. The following subsection will map out the process of PFI/PPP emergence in the UK. The development of PFI over the past 25 years is mapped out in Figure 1.1.

PFI/PPP post-1980

During the 1980s, the provision of private finance for public projects was governed by the Ryrie Rules (Ryrie was a Treasury official). The Ryrie Rules set two main criteria for the use of private finance in the procurement of public sector capital projects. These are:

1. Public sector capital projects can only be financed by the private sector if it is proved to be more cost–effective than a comparable publicly funded project.
2. The investment or asset must be part of the planned public sector schemes and not in addition to (in some cases, this rule is ignored by decision-makers). The consequences of this second rule are that public expenditures ought to be reduced by the same value as private investments.

Only a limited number of projects have been conceived under these two rules. Among these projects include the third crossing of the Thames at Dartford and the second Severn crossing. Many authors and politicians have commented that these rules have been a huge obstacle to encourage public authorities to seek private funding. This led to the abolishment of the second rule in 1989 by the UK Treasury which had the effect that privately funded projects no longer

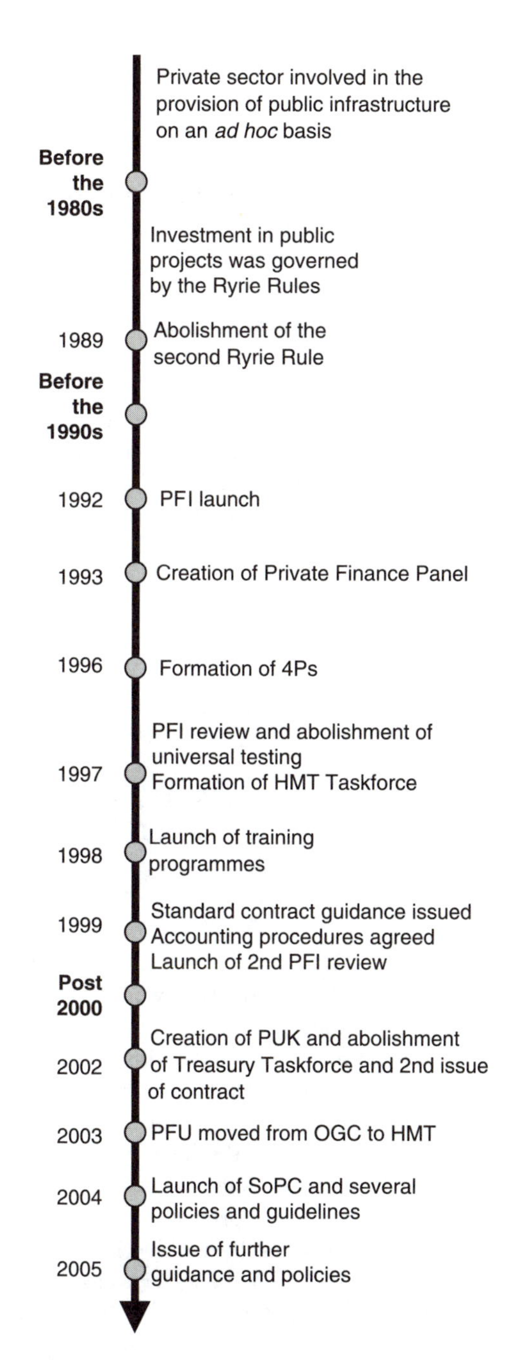

Figure 1.1 Map of PFI/PPP development.

had to be substitutional. Finally, both rules were discarded in 1992 with the introduction of the PFI/PPP in the procurement of public services.

PFI/PPP post-1990

In the early 1990s, the UK economy suffered a considerable slowdown. This reflected on the ability of the government to raise sufficient tax revenue to adequately improve the public sector's infrastructure and services. This directly or indirectly led to the creation of new legislation for the introduction of private finance to develop and operate public services and assets:

> The cumulative effect of under-investment on the capital stock inevitably increases through time. It was the reality of this pressing infrastructure need, alongside the equally pressing requirement to keep public expenditure under control, which, when coupled with an ideological commitment to involve the private sector in the public sector, led the Conservative Government to launch PFI in 1992.
>
> (Broadbent and Haslam 2000)

PFI was launched in 1992 by the Chancellor of the Exchequer. This launch was met with little enthusiasm by the private sector. As a result, in 1993, the Chancellor gave the PFI greater emphasis and momentum. He announced the creation of a Private Finance Panel. The main aim of this panel is:

> To encourage greater participation in the initiative by both private and public sectors, to stimulate new ideas, to identify new areas of public sector activity where the private sector could get involved, and to seek solutions to problems which might impede progress.
>
> (Private Finance Panel 1995)

In 1994, the Chancellor insisted that the Treasury would not approve any capital project unless the options to acquire private funding had been explored by the procuring department. This universal testing decision had the effect that public sector authorities are forced to engage with the private sector. This decision has played a pivotal role in securing private funding for the development of services and assets in the public sector. In 1996, the local association established the Public–Private Partnerships Programme (4Ps). The aim of 4Ps is to assist local authorities and encourage the growth of private investment in public services. Even with these initiatives, the uptake of PFI/PPP by the private sector fell short of the government's hopes. In 1997, after the general election, the new Labour Government re-affirmed its commitment to the PFI/PPP paradigm and appointed a committee to conduct a wide-ranging review of PFI/PPP. The new government also abandoned the universal testing of private sector financing in May 1997. The review committee made several key recommendations. Among these were:

- abolishment of the Private Finance Panel;
- control of PFI by the HM Treasury;
- compulsory not to shortlist more than four bidders;
- formation of a new private taskforce. The advice is that this entity should have a limited life and provide both strategic policies on PFI and technical advice on design and approval of specific projects.

These key recommendations led to centralisation and standardisation of practices across procuring departments. As a consequence of the review and formation of the Taskforce, the following have been achieved:

- 1998 The step-by-step guide to the PFI procurement process was launched.
- 1998 Development of a comprehensive approach to PFI training. A training programme on how to undertake PFI schemes was put in place.
- 1999 Standard contract guidance was issued.
- 1999 Agreement on accounting procedures for PFI schemes was reached.
- 1999 Establishment of a review group to sign off all local authority PFI schemes.

Other major developments in 1999 were the announcement and launch of a second review on PFI/PPP by the government. The main recommendations from the second review are:

- Formation of an advisery body composed of private sector experts to replace the technical advisery wing of the Treasury Taskforce.
- All PFI standard documentation has to be agreed by the Treasury Taskforce.
- Procuring departments are no longer encouraged to develop their own procedures.
- The project review mechanisms (sign-off gateways) are to apply to all procuring departments.
- Only risk associated with the operational aspects of building is considered for transfer.
- It is not a requirement that staff employed in a PFI deal need to be transferred to the private sector.
- Emphasis on value for money and the use of public sector comparators in making investment decisions.

As can be seen from the above list of recommendations, the 1990s era of PFI/PPP was dominated by standardisation of processes, procedures and emphasis on value for money in the delivery of public services and assets.

PFI/PPP post-2000

In this period we saw the consolidation of PFI/PPP policies on several fronts. In 2000, the Treasury Taskforce Policy section became part of the government's Office of Government Commerce (OGC), while the project's technical advisery team were transferred to a new entity, Partnerships UK (PUK). PUK is a PPP organisation, in which the government has a 49 per cent shareholding and the private sector has the other 51 per cent. In 2001, the HM Treasury experimented with a scheme under which ancillary staff, in soft facility management areas, were managed by the private sector but remained public sector employees. In 2003 and 2004, the HM Treasury issued a new publication *Standardisation of PFI Contracts* (SoPC), standardising the contract terms and related standard terms. *The Green Book* (HM Treasury 2003) introduced new procedures to ensure that all new PFI/PPP projects were to undergo a comprehensive evaluation after completion of the construction phase or at some later stage in the project life cycle. Existing projects are subject to retrospective evaluation. For example, the Department of Health has issued a *Good Practice Guide: Learning Lessons from Post-Project Evaluation* to assist health scheme managers to assess the impact of a project, programme or policy while it is in operation, or after it has come to an end. Also, in 2004, NHS Estates issued a design and development protocol for PFI schemes. Similarly, in 2005, the Housing Corporation produced a guide on good practice for housing PFI. Since its inception, 4Ps has released several guidelines and reports on PFI/PPP (for further details on their reports, consult their website). OGC has been very busy producing policies and strategies on public sector procurement, ranging from value for money to a checklist of guidance publications. The NAO also has undertaken several studies to assess the effectiveness of partnership and value for money in PFI projects.

In summary, the 2000s emphasised the enforcement of the standardisation of PFI contracts across both the public and private sectors. At the same time, the HM Treasury encouraged procuring bodies to develop SoPC-compliant sector-specific contracts. Several public and private organisations emerged to give support to public procuring departments and advice to the private sector organisations. This era's agenda is also dominated by risk transfer and value for money concerns.

Economical rationale for PFI

During its evolution the PFI/PPP has had both its supporters and critics. Much of the debate over the economic rationale for or against PFI/PPP has come from the two opposite ends of the spectrum. The first group is those who believe strongly that PFI/PPP has merit in providing public services. The second group is those who hold the view that there is an overwhelming economic case against the PFI/PPP procurement paradigm. The following subsections will present the arguments for and against PFI/PPP based on economics, risk transfer and value for money drivers.

The economic case against the PFI

It is clear from some of the definitions in the previous sections that one of the driving forces in the introduction of PFI procurement is related to problems over the lack of public finances to fund public expenditures. Critics of the PFI have argued that it is cheaper for the public sector to raise capital from the market than the private sector. This view is accepted by the HM Treasury (2000a) who admit that the private sector incurred higher costs in raising capital. It is estimated that the capital cost for the private sector is between 1–3 per cent higher than for the public sector. This fact has led many authors and observers to cast doubt on the economic claim that the PFI/PPP is a better model for delivering value for money than traditional public sector financing methods. In general, the economic case against the PFI/PPP model consists of the following criticisms:

- The appropriateness of the capital budgeting method used in option comparisons. Shaoul (2005) stated: 'Recent developments in finance theory throw increasing doubt on the appropriateness of the NPV rule in capital budgeting since the cash flows from a project cannot readily be forecast beforehand and argue that much of the value consists of the options that are opened up once a project has started: real options pricing models.'
- PFI finance is more expensive than conventional public investment finance (Spackman 2002).
- The claims that the PFI projects are more cost-effective than conventional public sector projects have been contested by several authors such as Sussex (2002), Pollock *et al.* (2002), and UNISON (2004).
- Excessive adjustments for bias (overstating benefits, and understating timings and costs, both capital and operational).
- The suggestion that the PFI/PPP model is based on a sound fiscal theory is vigorously challenged by Pollock *et al.* (2001): 'There is no economic or fiscal case for PFI and PPPs, but the Treasury continues to force public agencies down the private route.'
- The high return on capital that the private sector receive from PFI/PPP schemes (Ball *et al.* 2000; Gaffney *et al.* 1999).
- PFI/PPP procurement systems are very complex and involve a high level of transaction costs. These costs are borne by the public sector (Froud 2002).
- The relationship between the borrowings associated with the PFI, the government's balance sheet, and the size of the budget deficit. It is argued that the budget deficit is lower than it would otherwise have been if expenditure

on traditional public sector investment had taken place (Broadbent and Haslam 2000).

- Lack of a level playing field when considering PFI/PPP. Pollock *et al.* (2001) argued: 'The government continues to force public agencies down the partnership route. In many cases, adequate public funding is simply not available and public agencies are forced to pursue larger schemes that are more attractive to the private sector.'
- Scepticism about value for money. Authors such as Shaoul (2005), Froud (2002) and Pollock *et al.* (2001) strongly challenge the paradigm that the public sector can gain a high level of value for money from the managerial skills of the private sector that will result in greater efficiency and economy of running public services and projects.

The economic case for the PFI

Those who support the PFI/PPP financing model base their argument on macro-economic and micro-economic reasons. At the macro-level, it is pointed out that the Treasury's fiscal rules (the golden rule – the government will borrow only to invest and not to fund current spending, and the sustainable investment rule – debt-GDP ratios will be held at a stable level) could easily be adhered to without a finance injection from the private sector. Hence, PFI/PPP is not about getting capital spending off the balance sheet, but is about delivering value for money in a wider context of modernisation of public services (Robinson 2000). The Treasury Taskforce HM Treasury (1999a) has also reinforced this concept, stating:

> The PFI is not about borrowing money from the private sector . . . It is all about creating a structure in which improved value for money is achieved through private sector innovation and management skills delivering significant performance improvement and efficiency savings.

This opinion is supported by Hawksworth (2000) who emphasised: 'there is no obvious macro-level constraint on public sector investment that is driving the use of PPPs. This is true now and for the foreseeable future.'

Morley (2002) listed the benefits that may be gained from PFI/PPP procurement:

- fully developed business case for the project;
- appropriate allocation of the risks;
- proper consideration of all procurement options;
- competitive bidding for contractor selection;
- innovation in project design and construction;
- value for money testing of contractor proposals;
- change mechanisms embedded in the contract;

- collaborative approach to project implementation;
- development of longer-term relationships;
- reduction of political risk;
- opportunity to develop and enhance user choice.

Those who are in favour of the PFI/PPP economic model argue that PFI provides additional investment for the public sector and enables projects to proceed which would not otherwise do so (NAO 2001). This observation is supported by Bell (2005) who explained that 'PPP has arguably achieved what it set out to deliver, increased spending on public infrastructure, on a basis that provides predictable costs to the public purse and transfers major risks.' This view is also sustained by Stewart (2005). He states that the PFI model has a strong record in the delivery of public sector infrastructure projects on time and on budget. At the micro-level, research has also demonstrated that savings on PFI projects vary from 3 to 12 per cent (NAO 2001). Arthur Anderson and Enterprise LSE (2000) showed that an average saving of 17 per cent against PSCs was achieved on a sample of 29 projects. It is also pointed out that before PFI/PPP, 70 per cent of central government construction projects suffered from time and cost overruns, but since the introduction of PFI, the figure has dropped to 20 per cent (NAO 2003a). It also observed that because of the nature of PFI/PPP contracts, the public sector does not have to make any payments until the asset is delivered and operational. A comparison between PFI/PPP and traditional procurements is shown in Figures 1.2 and 1.3. The figures also illustrate that operational expenditures are based on a uniform unitary payment which helps to reduce the fluctuation of operational costs. These issues have been cited as an indication that PFI/PPP projects are far better at keeping to time and budget than other forms of procurement. Thus, PFI/PPP projects generate savings to the public

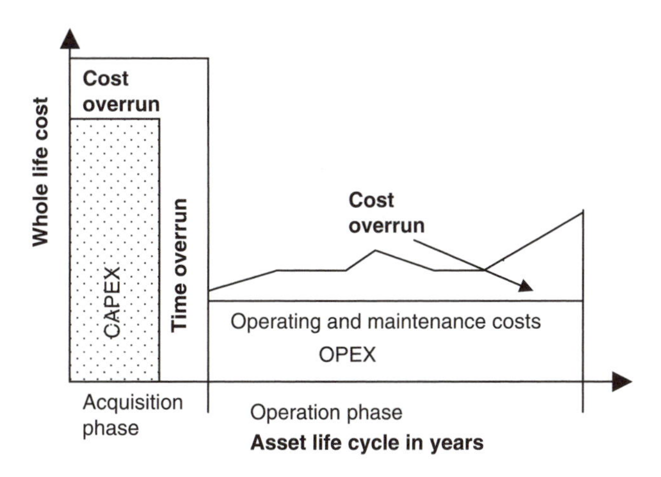

Figure 1.2 Traditional public procurement of building assets (adapted from PricewaterhouseCoopers 2003).

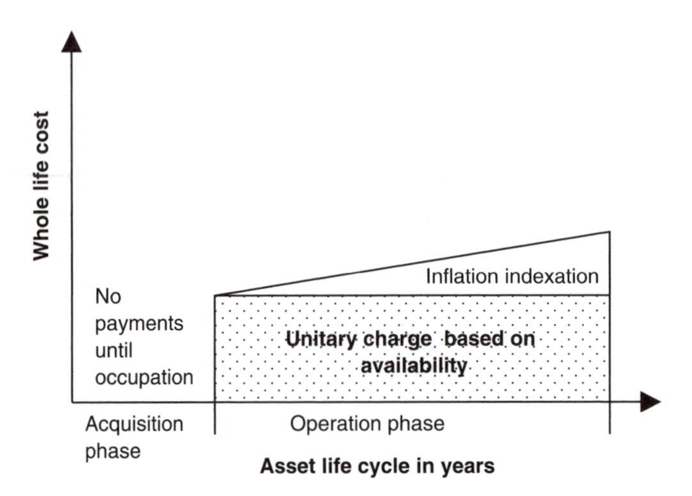

Figure 1.3 PFI/PPP procurement of building assets (adapted from PricewaterhouseCoopers 2003).

sector during the construction stage. This observation is sustained by Kent (2003) in the following statement:

> The procurement of the physical assets has been an unqualified success. All seven prisons have opened on time or early and all have opened to cost. They were delivered in about half the time taken by prison projects under previous routes of public procurement. They were also delivered at a far lower cost, and the PFI route led to lower capital costs than 10 years ago.

Other points of view suggest that it is the reduction in operational expenditure through increased capital investment that is unique under the PFI/PPP model. This view is based on the premise that the length of the concession agreement will enable providers to increase capital expenditure on building assets. It is speculated that the additional capital spent will be recovered through savings in operational expenditure over the duration of the project agreement.

Further details on the economic merits of project financing to fund public projects are presented in Chapters 4 and 5.

Value for money rationale

The PFI/PPP procurement model is based on the paradigm that all projects must generate value for money (VFM) for the public sector. The question of value for money is inherently linked to the issues discussed in the previous sections and, more importantly, to the risk transfer. The value for money subject is not discussed in depth here. Because of its paramount importance Chapters 2 and 3 are devoted to presenting a new theoretical framework for redefining the whole life cycle value of building assets. The theory is based on value creation and exchange throughout the life cycle of building facilities.

It is a requirement that the procuring authorities have to show that a PFI/PPP project generates VFM to the public sector. VFM should represent the optimum combination of costs over the life of the project, relative to the benefits provided. VFM analysis is carried out by looking at alternative options for acquiring the same project. VFM is based on the present value of these alternative options. HM Treasury guidelines emphasised that procuring departments must make real choices between competing procurement routes. The option that demonstrates the lowest present value is normally selected, because it represents the lower cost to the public sector. This approach has been criticised by Grout (1997). He demonstrated that the VFM is predominately biased against private sector finance. Broadbent and Laughlim (2000) pointed out that 'the issue of value-for-money has been the subject of extensive criticism in terms of whether it is possible to achieve sufficient savings to outweigh the basic fact that privately borrowed capital is more expensive than government borrowing'. However, research by PricewaterhouseCoopers (2002) for the Office of Government Commerce (OGC) found that the 'average spread between the project IRR (internal rate of return) and benchmark WACCs (Weighted Average Cost of Capital) has been some 2.4 per cent in total'. This issue is further complicated by the Treasury Taskforce (HM Treasury 1999a) who stated that 'PSCs alone are not the basis for assessing value for money'. These findings confirm the result of a review by Arthur Anderson and Enterprise LSE (2001). The review found an estimated average savings against PSCs of 17 per cent, as well as a narrowing gap between the cost of private sector capital and public borrowing. Others argued that '*real value lies not in the cost of the building work itself, nor in the cost of running the building, but rather in putting the facility to work for its intended purpose – in short, value lies in the outcome*' (The Smith Institute 2005). This demonstrated the need for a continuous assessment of VFM and development of key benchmark performance against realistic PSCs. Probably it is too soon to judge whether PFI/PPP projects have provided VFM to the taxpayers.

Risk transfer rationale

The PFI/PPP procurement model is intended to optimise the allocation of risk between the public and private sector. The PFI/PPP paradigm is based on the fact that the risk should be transferred to the party best able to manage it. One of the contentious issues in risk transfer in PFI/PPP contracts is how to assess and price risks. The measurement and pricing of risk are complex and problematic and yet play a significant part in the judgement between the PSC and the PFI/PPP alternatives selection. Chapters 12 and 13 provide further in-depth analysis on risk allocation and pricing in PFI/PPP projects.

It is argued by NAO (2003a) that PFI/PPP contracts provide greater time and cost certainty and hence show evidence of the successful transfer of risk to the private sector. According to NAO, 76 and 78 per cent of projects procured under PFI/PPP have been constructed on time and to budget, respectively. In contrast, only 30 and 27 per cent of projects procured under conventional procurement have been constructed on time and to budget, respectively. This success could be attributed to other factors rather than risk transfer. It is important to point out that these savings are asset-related rather than savings in the provisions of services. Further studies may be required to confirm the NAO findings and demonstrate that savings can also be made in the provisions of services. Edwards and Shaoul (2002) concur with this view:

> Our analysis has shown that the concept of risk transfer that lies at the heart of the rationale for partnerships is problematic, regardless of whether the project is 'successful' or not. If the project is successful, then the public agency may pay more than under conventional procurement: if it is unsuccessful then the risks and costs are dispersed in unexpected ways. Hence public accountability is obscured . . . our analysis shows that, although a project fails to transfer risk and deliver value for money in the way that the public agency anticipated, the possibility of enforcing the arrangements and/or dissolving the partnership is in practice severely circumscribed for both legal and operational reasons.

Despite these concerns, others have argued that the transfer of risk to the private sector provides an incentive to the private sector to maximise efficiency and thus deliver value for money to the public sector (PricewaterhouseCoopers 2002). The issue of risk pricing and transfer needs further investigation before we can conclude that the efficiency seen in some PFI/PPP projects is attributed purely to risk allocation to the private sector.

The benefits of public–private partnerships

The key benefits from the utilisation of PPP/PFI procurement have been pointed out, including the transfer of construction risks (i.e., time and cost overruns), fixed payments over the life of the asset, as well as the link between the quality of service provided and the actual amount of payment to the private sector (House of Commons 2002). In a study on the experience of the use of public–private partnerships in the UK, PricewaterhouseCoopers (2003) explained: 'While there may be an additional financing cost for the use of private sector funding, this will in many cases be offset by the synergies gained from combining design, construction and operation.'

It has been claimed that one of the overriding successes is the ability of private and public partnerships to deliver a substantial public investment on time and to budget. Other benefits arising from public–private partnerships include:

- adaptation of whole life cycle costing;
- integration of design, construction and operation of building assets;
- effective partnership with the private sector;
- achieving long-term value for money;
- aligning the interests of private and public sectors.

These are some of the benefits cited for public–private partnerships, but certainly the jury is still out, and we will not be able to state categorically that the public–private partnerships have delivered value for money to the taxpayers for another decade or two.

Future PFI/PPP directions

There are still large gaps in the PFI/PPP knowledge base and its performance measurement in practice. It is anticipated that the development of PFI/PPP in the coming years will involve consolidating some of the existing practices and ensuring the long-term variability of PFI/PPP procurement as an alternative, offering better solutions for construction and value for money to the taxpayers. According to the Smith Institute (2005), the issues that need to be addressed include:

- defining value for money;
- flexibility of PFI/PPP deals;
- post-construction project appraisal;
- lower bid costs to encourage more contractors to compete for projects;
- how can the public sector ensure that PFI/PPP can deliver long-term sustainable value?
- flexibility of contract to accommodate the long-term necessary changes;
- more refinement and standardisation of contracts and procedures;
- relaxation of procurement rules to allow for true alliancing and partnering principles to be implemented in PFI/PPP deals;
- review of risk transfer and the inequities that may exist in the present PFI/PPP deals;
- focusing on outcomes rather than output specification;
- improving the public sector's skill capability;
- development of PFI/PPP bids with a stated benchmark cost figure;
- development of an efficient relationship management;
- embracing PFI/PPP decisions based on value rather than cost;
- development and implementation of electronic procurement systems;
- improvement of design quality of PFI/PPP projects;
- streamlining the bidding process;
- developing the knowledge base and mechanisms to share learning from successful approaches/projects;

PFI/PPP as a driver for whole life cycle costing

The HM Treasury has demanded high certainty about the true costs and benefits of new building assets over the entire life cycle of building assets. This has led to the emergence of whole life cycle costing. Whole life cycle costing is seen as the most reliable method for determining the cost-effectiveness of PFI/PPP projects. This view is supported by the OGC in their Procurement Guide 7 (2003b):

> Value for money is the optimum combination of whole-life cost and quality to meet the use's requirement. This means that awarding contracts on the bids of lowest price tendered for construction works is rarely value for money; long-term value over the life of the asset is a much more reliable indicator. It is the relationship between long-term costs and the benefits achieved by clients that represents value for money.

The OGC statement directly links whole life cycle to value for money and explicitly emphasises that it is the relationship between whole life costs and benefits that can be generated from a PFI/PPP deal that constitutes value for money. The challenge for both the private and public sectors is to minimise whole life cycle costs. The ultimate objective is to find an optimum strategy to trade off a higher capital cost for lower recurring operating and maintenance costs. This trade-off is especially difficult to achieve over a 30-year contract period. However, Arthur Anderson and Enterprise LSE (2000) found that 'the private sector appears best able to generate value for money savings where projects have a high proportion of capital expenditure'. It is argued that one of the benefits of PFI/PPP is the motivation by private finance to bidders to use whole life costing. PricewaterhouseCoopers (2003) stated:

> Private finance and operation will usually avoid the cost and timetable slippages that have been common under traditional public procurement. This approach encourages bidders to focus on the whole life costs of the asset over the project life cycle because those responsible for the building of an asset are also responsible for long-term maintenance and operation.

The UK Government has issued guidelines to all procuring authorities to ensure that all procured new buildings must demonstrate value for money over the long term, and not to concentrate on the design solution which is the least expensive. These changes have led to and highlighted the importance of whole life cycle costing approaches to the design, construction and operation of buildings. As a matter of policy, all public investment and disposal decisions over a certain level will require an option appraisal and life cycle costing assessment. This insistence from the largest client in the country has empowered contractors, developers, funders, private clients, building insurers, service providers, manufacturers, etc., to seek whole life solutions.

The challenge of cost planning PFI/PPP projects

The PFI/PPP procurement process requires cost analysis and management from cost experts who will be responsible for developing cost plans, costing design, life cycle replacement, and pricing all aspects of the asset. The costing environment surrounding PFI/PPP projects is significantly different from conventionally procured projects. The cost planning process requires cost analysts to use their expertise and innovative thinking to develop whole life cost solutions that deliver value for money to the client, and thus improve public building assets performance. Over the last few years, we have seen the emergence of much greater complexity in the costing of PFI/PPP projects, complexity in terms of process, scope, number of items to be priced and specified, and the level of detail required for whole life cycle costing. Failure to understand and manage this costing complexity can result in extensive cost variations and a much greater risk of failure in delivering VFM for all stakeholders concerned.

One of the challenging aspects of cost planning in PFI/PPP projects is the provision of forecasts over different time horizons. The question that needs to be asked here is, how can a cost analyst reconcile the concession period (a time horizon of over 30 years) and the dynamisms of client expectations over this time horizon? The problem here is that most cost analysts are trained to price inputs rather than output specification. The challenge also comes from the forecasts of future income streams, life cycle funds, the life service of individual components, operational costs, the timescale of life cycle replacement, and other similar financial forecasts that are required in PFI/PPP procurement. The challenge for the cost analyst is to convince all stakeholders that these are only forecasts and should never be seen as being guarantees of future cost outcomes. Other challenges that may face cost analysts are the impact of technological changes on the rate of changing or updating building components and equipment and making long-term cost forecasts on minimal information. Little information is available about the service life of building elements, and the cost analyst is not trained to price against output specifications. Cost analysts should recognise that changes in costing assumptions are inevitable, changes will be necessary to take into consideration user demand, technological, political, and economic and policy uncertainty. The question that needs to be asked is, how can a cost analyst price for uncertainty? Currently, long-term price uncertainty is dealt with through contract clauses, contingency allowances and inflation indexation. Maybe the way forward is to only price for a short period of time, e.g., five years, as well as allowing the review of the contract prices (not only the unitary payment but other costing elements as well) at specified periodical intervals. This enables cost analysts to develop long-term forecast range values, and short-term prices with high certainty. Other methods for managing cost uncertainty include cost-reimbursement pricing, profit and cost sharing and performance-based pricing.

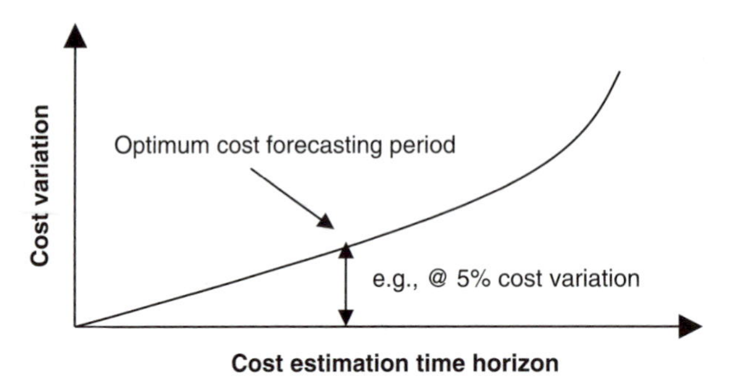

Figure 1.4 Relationship between cost variation and period of forecasting.

If cost and time overruns and variations are still common through the construction phase of PFI/PPP projects, how much more inevitable is it that cost change will be required throughout the 30 years of the project agreement? There are limits as to how much certainty a cost analyst can bring to a cost plan that has a time horizon of over 30 years. It is impossible to know cost projections over a 30-year timescale, and it is unwise to develop cost plans even though cost analysts may attempt to. Price variation and certainty depend on both the length of the forecast, the level of risk, and the complexity of the PFI/PPP deal. As demonstrated in Figure 1.4, the longer the forecasting period, the more likely cost variation will occur. Thus, costing assumptions and decisions should be based on the knowledge that prices are likely to be between a certain range of values rather than any precise price that might emerge from the cost planning process. For long-term contracts, it is better to formulate a range of prices or, even better, a price profile for each cost element. In this way, cost analysts and decision-makers can easily manage the cost movement of any particular contract within the established range boundaries.

The challenges stated in this section will serve as the guiding principles for this book. Chapters 6–11 address the cost planning aspects of PFI/PPP building projects. The aim is to convey to the reader the fundamental principles of cost planning and the challenges that face the process of pricing a PFI/PPP contract.

Summary

In summary, this chapter has introduced the rationale for and against the PFI/PPP paradigm. Those who are against believe that the lack of a level playing field between PSCs and private options render the value for money and risk transfer judgements more biased to the private sector. Those who are in favour quote the benefits achieved so far from public–private partnerships in terms of delivering large public sector investment on time and to budget. The chapter has also stated the importance of cost planning PFI/PPP projects and the changes that this may pose to cost analysts. Cost analysts are asked to price contracts for the long-term horizon with little information on most of the building elements and operational aspects of the services. Cost analysts are trained to price inputs, not output specifications. This is a challenge that we must address as a whole industry.

Chapter 2

Whole life cycle value exchange

Introduction

The intensity of the value debate in the built environment disciplines has generated significant interest in investment appraisal methodologies such as developing whole life cycle costs (WLCC). The advocacy of holistic approaches to built asset decision-making has existed for many years (Boussabaine and Kirkham 2004) but the recent WLCC paradigm has perhaps been more influential than others in driving the value debate forward. The literature reveals evidence to this effect (Pearce 2003; OGC 2003b) and attempts to infer value from whole life-based ratios (Royal Academy of Engineering 1999) have undoubtedly invigorated a great deal of interest (and dispute) in the measurement of 'value' within the built environment.

The result is that researchers are now asked the question, 'What do such ratios tell us, if anything, about value?' While it could be argued that these ratios demonstrate the importance of considering the whole life cycle costs (WLCC) of an asset, the mere fact that the numbers used to calculate these ratios are questionable lends little support to that notion. Clearly, the following inferences, among others, *cannot* be drawn from them:

- that comparative information on different buildings will allow users to ascertain if one building is more economical than another;
- that these ratios can be used to determine cost control effectiveness both at design and operational stage;
- that they can demonstrate that value for money, in terms of business (or service) performance, creation of value, etc. has been achieved in a built asset;
- that ratio can be used to test the effectiveness of design in meeting functional and performance requirements

In direct opposition to the ratio notion, the author contends that measurement of value cannot possibly be considered to be static, given the dynamic nature of the latent variables that contribute to and influence it. Indeed, the catalyst behind the decision to examine further the complexities of the value proposition has arisen as a result of the changing face of public service provision in the UK. The influences affecting traditionally procured assets will be discern; able from those associated with PFI/PPP schemes, for example. Therefore, any attempt to model value requires explicit knowledge of the effect of changes in stakeholder preference and need and political aversions as well as the micro-variables that will directly affect the asset.

In value engineering and value analysis, the concept of value is largely equated with reduced cost (e.g. Shillito and de Marle 1992). However, in UK construction industry context, Value management (VM) is seen as a more strategic, whole-life process, intended to 'maximise the benefits of a project, to the project's sponsor or client' (Kelly and Male 2002). Furthermore, very broad definitions of what constitutes inputs to and benefits from a project are adopted, certainly extending well beyond simple financial considerations. It is generally held that while cost reduction is, in some cases, a consequence of VM, it is not its principal objective (McGeorge and Palmer 2002). Nonetheless, while it is commonly acknowledged that each of the many stakeholders in a construction project will have different perspectives on what constitutes 'value', it is clear that conventional VM focuses on the client's value system (see e.g. Kelly and Male 2002). One of the key aims of the work described in this chapter is to begin to establish a value framework that will enable optimisation of value across the entire range of project stakeholders. Palmer *et al.* (1996) investigated the theory and practice of value engineering in the construction industry in the USA. They concluded that value engineering in USA is mainly a design audit process. The work of Chen and Liu (2003) identified key factors for value management success. Their findings confirm that value management studies require a combined effort from all parties involved. Dikmen *et al.* (2005) developed a conceptual framework to investigate value innovations within construction companies. Cha and O'Connor (2005) also developed a tool that assists in achieving project value objectives.

Thomson *et al.* (2003) provide a review on value and quality attributes and their relationship to design activities. Lapierre (1997) discussed the concept of business-to-business professional service value and developed a model that illustrates how value definition is a function of time. It has also been emphasised that value exchange relationship activities add value to a service (Crosby *et al.* 1990).

Blois (2003) points out that there are two sides to an exchange. Parties entering into an exchange will have a list of tangible and intangible benefits to be negotiated in the hope that the ownership of these perceived benefits will increase the chances of creating value for all the parties involved. Miller and Lewis (1991) argue that the foundation for value exchange theory exists in recognising values or value systems as fundamental to all individual or organisational behaviour. Groth *et al.* (1996) stated that companies should exist to create value to themselves and their customers. Bowman and Ambrosini (2000) suggest distinguishing between use value and exchange value. They define use value as the specific qualities of the product or service perceived by customers in relation to their needs, whereas exchange value refers to price or amount of cost realised at a single point in the whole life cycle process when the exchange of services or products take place. The authors also pointed out that products and services at the point of exchange have both an exchange value and a perceived use value. They also remarked that exchange value is not transferred to the organisation's production or distribution process, only use value is.

Considering that WLC value exchange mechanisms are complex, inherently risky and may have a cumulative effect on adding value to products/services, an investigation into the concept of WLC value exchange, which includes all the principal actors is overdue and essential in order to complete the picture in this specific field.

Demonstrating the application of WLC approaches to the value proposition requires an appreciation of the dynamic nature of the characteristics of built asset stakeholders, principally from an economic and chronological perspective. The ambiguous nature of the value concept must also be fully appreciated in order to consider the differences in value definitions. The chapter illustrates the concept of 'value exchange' in the WLC, 'convergence of WLC stakeholders' value', followed by a discussion and consideration of the research implications.

Value overview

In recent years there has been an increased interest in the built environment sector in the concept of value exchange and value theories in general. There is no question that both exchange theory and value theory are central concepts in procuring built environment assets within the whole life cycle framework. Yet they are not appreciated by the sector professionals, especially in the context of the exchange of services, where value exchange has not yet been defined and researched to any great extent.

The role of WLC stakeholders is to search for management mechanisms that will maximise value exchange. In order to appreciate the efficiency and effectiveness of WLC management processes, WLC stakeholders require a clear understanding of value concepts and their emergence as a performance metric. A theoretical framework of WLC value exchange is essential before passing judgement on whether whole life processes are delivering or adding best value to stakeholders' organisations, the end-user, the environment and society as a whole.

One can argue that WLC value is neither simply measured by price, nor is it the WLCC budget. Indeed, many consider that value is mainly a subjective concept (e.g. Jackson 2001). Thomson *et al.* (2003) stated that there is a need for value language and proposed definitions for quality and value attributes. Bourguignon (2005), for example, stated that value encompasses a multitude of meanings that can be divided into three clusters: 'philosophical', 'measurement' and 'economic'. We will also briefly examine the related concept of 'value for money' and the process of value management (VM).

Philosophical value

It is well recognised that many theories (e.g., motivation, leadership, etc.) are connected to value(s) in all forms including aspects relating to conflict, relationships, stress, personal ethics, and values (Miller and Lewis 1991). Philosophical value can be seen in terms of both subjective and objective values. Subjective philosophical value is related to aspects of truth, justice, equity, trust, norms, etc. and it is also related to value judgement. Objective philosophical value can be regarded as an extension of the economic usage value, based on social utility theory and its instrumentals aspects are related to the ability to satisfy the needs or functions (Bourguignon 2005). All these aspects are an integral part of modern procurement systems that are based on trust and equity for every WLC stakeholder, for example, 'Partnering'.

WLC philosophical value may be defined in terms of the morals or morality of any system of social values and rules of conduct prevailing within participating organisations. WLC stakeholders pay lip service to norms such as 'loyalty' and 'trust'. The concept here is to combine each WLC firm's individual action and collective action, as well as the conditions of efficient action to deliver value based on ethical codes such as loyalty, trust, etc.

Measurement value

According to Bourguignon (2005), value is defined by measurement and measurement by value. The concept of measurement is related to quantification of subjective aspects, such as a sense of value, and objective aspects, such as mathematical and physical aspects of the real world.

Value assessment and measurement are necessary components of any WLC decision-making process. Value measurement can be defined as the accurate measurement of the cost effectiveness of WLC processes and products. Cost effectiveness here is related to the assessment of WLC stakeholders' value through utility elicitation or usage of services and procedures. Utility, in this context, is the price representing the value WLC participants place on services or products.

Economic value

Economic value represents the addition to value that can be measured in monetary terms (Groth *et al.* 1996). Economic value added (EVA) has been developed by the Stewart Corporation as an overall measure of financial performance that is intended to focus managers' efforts on delivery of shareholder value (Lovata and Costigan 2002). Since the development of EVA, many variants of value indicators have been suggested (Bourguignon 2005): market value added (MVA); shareholder value added (SVA); and total shareholder return (TSR). EVA is defined as accounting profit less a charge for capital employed, whereas MVA is defined as the additional value offered to investors by the financial market, i.e., market value minus capital (ibid.). It is argued that EVA is similar to residual income (Otley 1999). EVA takes a more historic view and only uses accounting rather economic valuations for the simple reason that it does not anticipate the earning of future income (ibid.). However, others argue that EVA is superior to other metrics because it requires managers to take a long-term planning horizon and gives managers clearer signals as to how to increase shareholder values (Lovata and Costigan 2002). This may cause a problem in measuring whole life value exchange since most of the analysis is related to long-term future income rather than past performance. Having said this, EVA pays particular attention to the setting of appropriate targets. This may help in setting benchmarks targets at each stage of the WLC process, especially financial ones.

Value for money

The concept of value for money (VFM) pervades a vast spectrum of disciplines, but is particularly prevalent in the key public sector disciplines such as healthcare and education and is concerned with the totality of the service provided rather than one specific aspect of it (Bannister 2001). In the UK, the term 'value for money' appears ambiguous – this is unsurprising given the contextual differences that occur in usage.

The UK's Office of Government Commerce (OGC) defines VFM as 'the optimum combination of whole-life costs and quality, to meet the user requirements' (OGC 2004b), whereas, Pollock *et al.* (2002) argue that value for money is based on a purely economic appraisal that compares the costs and benefits of alternative investment decisions. These authors describe value for money in the context of the private finance initiative (PFI), exemplifying the use of the public sector comparator (PSC), i.e. the notional annual costs were the scheme to be conventionally publicly financed, as a means of assessing VFM.

The dynamic nature of value for money assessment is important, particularly in the context of building procurement, and this was recognised in Griffith and Headly (1998). They argued that accurate and reliable feedback throughout is essential to ensure that value for money is being obtained. This theory can be observed in later practice-based publications for public sector procurement such as *Best Value* (ODPM 2005) and the *Gateway Reviews* (OGC 2004c). The authors also propose that value for money is determined by putting minimum resources to produce maximum benefit, although this argument is inconsistent with the dynamic nature of value for money that may require a more pragmatic approach to a basic optimisation strategy for resources.

Value management

The discipline of value management pervades many business scenarios, and has recently become more important in the global construction industry – particularly given the increased incidence of procurement routes such as BOOT/PPP. The definitions of value management are, however, variable, the Institute of Value Management and BS-EN 12973:2000 define it generically as a 'style of management particularly dedicated to motivating people, developing skills and promoting synergies and innovation, with the aim of maximizing the overall performance of an organization'. Interestingly, it is argued that VM is a generalisation for techniques such as value analysis, value engineering and value improvement (Langston 2005) although this is refuted by Neap and Celik (1999) who identify specific differences between value management and value engineering. The difficulties in implementing a consistent methodological framework for VM within a business setting are also recognised in Dumond (2000), who identifies that a lack of a universal implementation strategy across different organisations working together can lead to significant weaknesses. Some of the issues touched upon by Dumond were highlighted earlier by Simister and Green (1997), who identified stakeholder expectations, implementation, participation, power, time constraints and uncertainty as critical success factors in value management appraisals.

WLC value exchange

We adopt the definition of WLC value exchange as a tangible or intangible WLC product, service, knowledge, or benefit that is desirable or useful to its WLC recipients so that they are willing to return a fair price or exchange (Allee 2000). Accordingly, WLC parties could, for example, exchange knowledge for knowledge or product for knowledge or monetary value(s). The concept of WLC value exchange might be formulated in the following equations:

$$V_R = \frac{\sum_{j=1}^{j=n} S_j}{P}$$

$$V_C = \frac{\sum_{j=1}^{j=n} S_j}{C}$$

Where:

V_R = value received by any whole life cycle stakeholders.

S_j = satisfaction j in terms of tangible and intangible benefits

P = price (tangible and intangible) for value(s) received

V_C = value created for exchange by WLC stakeholders

C = cost incurred in creating V_C.

V_R and V_C are modelled as a ratio between satisfaction in terms of tangible and intangible benefits, and the price and costs for receiving and creating value. This will allow benchmarking between projects whereas measurement in absolute values will not. It is important to stress here that WLC tangible and intangible benefits will vary from project to project and will be different from stakeholder to stakeholder within any single project. This will constitute one of the major elements of future research into value and value exchange that this chapter proposes.

WLC value exchange is defined here as the search for optimum solutions that manage WLC relationships in order to maximise added value. This definition entails that WLC stakeholders must be able to evaluate the efficiency and effectiveness of WLC relationships and also must have a clear understanding of whole life value exchange mechanisms. This is crucial for the simple reason that the value added to the WLC process by each participating stakeholder should be more than the value of its inputs to the process, otherwise the value and cost are the same; and this may result in some WLC participant taking value from others. It is important here to stress that each WLC stakeholder must add value to their own resources as well as adding value to the other organisations they are serving. The management of a successful whole life cycle process must add value to all participating organisations. Hence, the urgent need for a whole life project management system that is able to organise, plan, control and coordinate WLC processes with the ultimate aim of adding maximum value to all participants. The overriding principle here is that value will be added by better management of WLC relationships. One can argue that the fundamental purpose of whole life management is to add value. The key element of WLC value exchange, that differentiates it from other traditional asset management processes, is the management of the whole life process on the value-based principles. As such, it must include the following (adapted from Malmi and Ikäheimo 2003):

- the aim to create WLC stakeholders' value;
- identification of the WLC value drivers;
- the connection of WLC performance measurement, target setting and rewards to WLC value creation or WLC value drivers;
- the connection of WLC key decisions and action planning, both strategic and operational, to WLC value creation and WLC value drivers (see Boussabaine and Kirkham 2004, Chapter 3);

Using the above principles for managing WLC relationships will have far-reaching consequences on the quality and operation of building assets. This will constitute a major shift from the traditional cost-based procurement to value-based procurement processes. If the above value principles are applied at every milestone of the building development and operation processes, it will allow the development of continuous systematic approach to value creation and exchange in the whole life process and go a long way towards achieving a sustainable added value creation process and ultimately to a sustainable built environment.

Figure 2.1 illustrates the continuous and iterative WLC value exchange relationship. The figure shows a potentially infinite chain of converting resources into products and products into resources. The purpose here is to show how value exchange revolves around the WLC participants. Value is created and exchanged constantly throughout the WLC process. The concept is that creating WLC value necessitates that each WLC stakeholder must listen to their clients and service users and respond with innovative solutions, policies, and procedures, thereby securing improvements in an efficient and effective manner. In other words, adding value to each WLC organisation through product and service developments and by ensuring that the value exchanges (services or product) that are supplied are what consumers (other stakeholders in the chain) demand. In utilising this concept, WLC stakeholders should act as though they are shareholders of the whole life process. Thus, the WLC value exchange model presented here can be used to measure both internal and external linkages in terms of the value given and the value received by all WLC parties.

Stakeholders' perspectives on value exchange

The following sections describe in detail the role of each WLC stakeholder in the WLC value exchange, as illustrated in Figure 2.2.

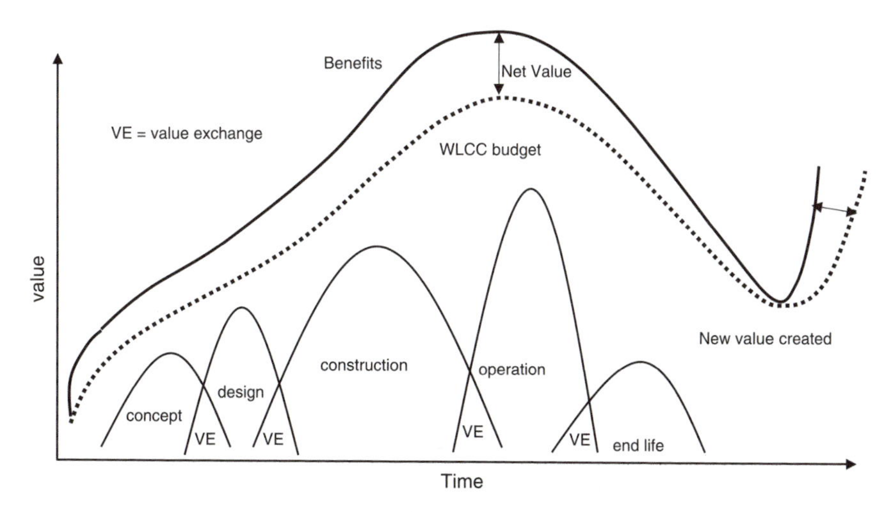

Figure 2.1 The concept of value creation/exchange in the WLC process.

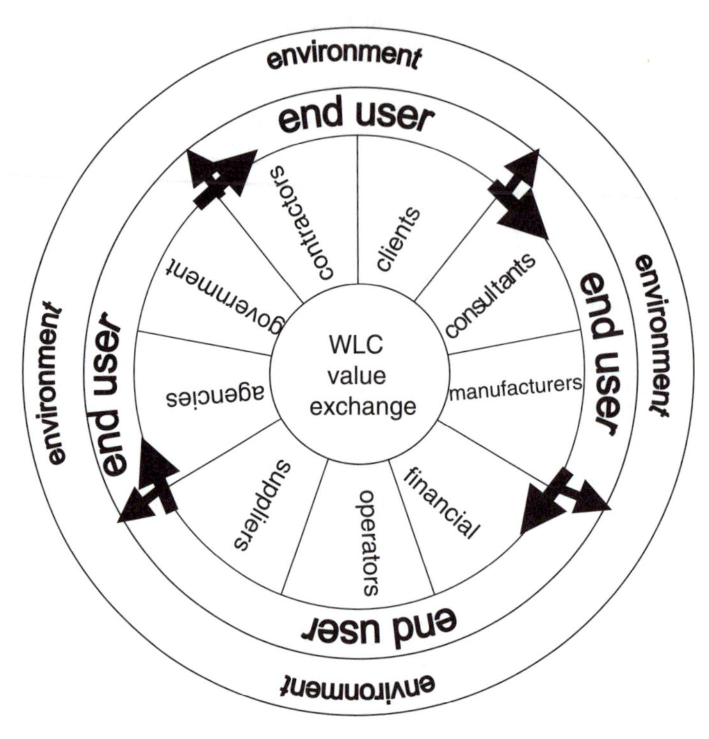

Figure 2.2 WLC value exchange stakeholders.

Clients' value

This category of value exchange has been most fully appreciated by WLC stakeholders as clients are demanding more value for their money. Clients are looking to add value to their investment in building assets through profitability, capital cost minimisation, WLC maintenance/operation costs minimisation, quality of service, time scale and early occupation, facility performance, durability, reliability of product and services, etc. Clients exchange a combination of nominal price and all other costs of product or service acquisition for the perceived benefits that are a combination of different attributes of product or service (tangible and intangible), available in relation to particular exchanges and use situations in the WLC process.

Financiers' value

Various theories exist that explain how and why financiers receive their share of value. However, the main thrust is that all suppliers of capital can capture a share of WLC value exchange through profit or other reward mechanisms, e.g. interest. It is important to stress here that the structure by which building assets are financed can have a significant effect on value captured by WLC stakeholders but to much less extent on non-financial aspects of the value creation process. Hence, it is prudent to select different financial packages to suit different projects based on the criteria of value creation not only to financiers but to all stakeholders on an equitable basis. Probably here there is a strong relationship between the nature of the financing method and the amount of value exchange that the capital supplier captures.

Suppliers' value

The increasing emphasis on supply chain management has highlighted the value added potential of this process. Through value-based management, the supplier works with key whole life stakeholders to improve interfacing processes, reduce operating expenses, and increase profitability, thus adding value to their shareholders/owners. The benefits of such practice can result in consequences that are beyond the simple exchange value of goods and services supplied, to include improvements in revenue, cost of construction materials and building asset costs. In fact, the effect of such a value-based supply approach can be traced to shareholders' value creation within the WLC customers' organisations. Also these supply value drivers can reduce the environmental burden through better optimisation and an interfacing process to minimise transport and waste.

Consultants' value

This value exchange is concerned with individual capabilities, knowledge, skills, experience and problem-solving abilities that reside in consultant organisations. Here, the human capital involved includes both physical and intellectual components. Intellectual capital or know-how conceives ideas, whereas physical capital converts ideas into reality. Throughout the WLC process, financial capital is converted into human capital and tangible resources. These resources are then used to provide services and produce building assets. Hence, the concept that some WLC stakeholders may exchange financial capital for intellectual capital. Consultants' knowledge capital reaches the desired ultimate form of productive capital when the knowledge is converted into a product or service capital that provides need fulfilment for any of the WLC stakeholders shown in Figure 2.2.

Government's value

This value is related to regulatory bodies and government groups. Government bodies create value through legislation in exchange for employment, growth and taxation. From the government's point of view, the value of the offered policies must exceed the price exchanges.

Manufacturers' value

A manufacturer must develop product offerings that add value to the WLC stakeholders in their terms. Such practice is more likely to result in the development of unique product offering than if the manufacturer develops a product offering assuming what he thinks value means to the WLC participant (Blois 2003). Hence, the concept that a manufacturer's value is not just the essential needs or values that the manufactured product being offered will fulfil, but is about how others in the WLC chain perceive the exchanged product contributing to generation of value within the WLC process.

End users' value

The end user value is concerned with customers' perceived value and can be applied to any of the organisations participating in the WLC chain. The utility, worth, benefits and quality theories, which lie at the foundation of the modern management and microeconomic theories, provide a theoretical underpinning for the end user value concept. These theories suggest that often end users do not buy products or services for their own sake. According to these theories, they exchange capital value for bundles of attributes whose value may be derived from the combination of attributes less the disutility represented by their sacrifices in obtaining the product or service (Snoj *et al.* 2004). This illustrates the complexity of the definition problem of end user value exchange. Indeed, this complexity of definition is stated and discussed in most end user value research. Snoj *et al.* provided a set of common denominators from different definitions of end user perceived value:

- value for a consumer is related to his expertise or knowledge of buying and using a product;
- value for a consumer is related to the perception of that consumer and cannot be objectively defined by an organisation;
- the customer's perceived value is a multidimensional concept;
- it presents a trade-off between benefits and sacrifices perceived by customers in a supplier's offering.

It has been suggested that 'customer value learning translation begins when a seller creates a value delivery strategy for target customers' (Woodruff 1997). These concepts of end user value, if adopted, will advance the practice of managing the WLC chain towards achieving end user value. The return of values from end user to WLC chain firms includes revenues, sales, creation of more investment and services opportunities, and many more tangible and intangible benefits.

Environmental value

The concept of environmental value exchange refers to the values that are imported to and exported from both the general environment and society as a whole. The environmental value is the value of WLC stakeholders' relationship with the earth and its resources as understood through true life cycle assessment (LCA) of the environmental impacts of resources consumed and pollution produced by WLC stakeholders. Societal value refers to the value of relationships and benefits enjoyed by society through the exercise of tangible and intangible value exchanges, including economic and social components. Social values would refer to the quality and value of relationships enjoyed with the larger society as a contributing member of local, regional and global communities (Allee 2000). Hence, it is beneficial for all those participating in the WLC chain to bring both society and the environment into the value exchange equation. It is possible that WLC stakeholders will be able to add value to their respective businesses while embracing management practices that are grounded in social responsibility and sustainable environmental practices.

Agency value

The concept of value addressed here is concerned with alliances and business relationships with all groups which have a stake in the built environment. WLC stakeholders' business relationships with all concerned agencies are valuable resources. Although they do not appear on the balance sheet, they can provide and add value to the services provided by WLC organisations. The relationship itself contributes to the total value exchange perceived by WLC firms through alliances and strategic partnerships. The concept of strategic alliances is vital and an important part of the atmosphere of the WLC value exchange.

Contractors' value

Contractors and subcontractors will have an exchange relationship with some of the WLC firms and, if there is more than one WLC organisation involved in the exchange, some form of composite exchange system will exist between the involved parties. Here contractors use resource capital value, such as people, equipment, materials and finance to transform the intellectual capital value (of design) into a building asset. Hence, contractors exchange capital resources for gross cash returns and other tangible and intangible benefits that cover the costs of construction plus the perceived benefits.

Operators' value

Operator value is related to both existing use value and value in use. Value in use is the value that a specific building asset has for a specific use to a specific user. The operator exchanges capital value, in terms of FM costs, replacement costs, maintenance costs, etc., for the contribution made by the building asset to the business of which it is part. Value in use tends to be synonymous with value to the entity, value to the business or value to owner (Sayce and Connellan 2002). Occupiers exchange total costs of occupying space for the perceived business benefits (value in use) in using the space.

Convergence of stakeholders' value

It is well recognised and documented that interpersonal and inter-organisational conflict, disputes and claims between WLC stakeholders have been the main contributors to adversarial relationships in the construction industry and have contributed to the imbalance in the value exchange throughout the life cycle of projects.

It has been assumed by many academics and practitioners, in recent years, that if all WLC stakeholders adopted ethical and equitable practices, in addition to sound WLC judgements, an increase of value would be realised for all involved parties. WLC values must be kept in essential equilibrium as a prerequisite for fair WLC value exchange. Furthermore, the avoidance of domination of any stakeholder's interests is very important to the attainment of value convergence. Failure to observe this balance can result in some stakeholders' failure to create value and potentially results in them taking value from others. These convergence problems can be remedied through the following efforts, modified from Miller and Lewis (1991):

- conscious management effort to maintain balance as a requisite to WLC value exchange;
- improvement of WLC relationships through a mutually beneficial increase in the exchange of values;
- continuous improvement in the type, quality of service and quantity of value received by all WLC stakeholders;
- WLC stakeholders directing their effort towards the combined elevation of all type of values (as defined above) toward a maximisation of the total set of value exchange;
- maximisation of WLC value through the integration and convergence of interest of WLC parties.

Figure 2.3 illustrates the principle behind value convergence between two WLC stakeholders. The figure shows that the value received by stakeholder (Y) or (X) should at least be equal or more than the value exchanged. The difference between the two is the net value. The greater the difference, the larger the imbalance in value exchange. This difference of imbalance in value exchange might be attributed to risks that each party is willing to take in exchanging value with other WLC partners in the life cycle chain. If the difference is

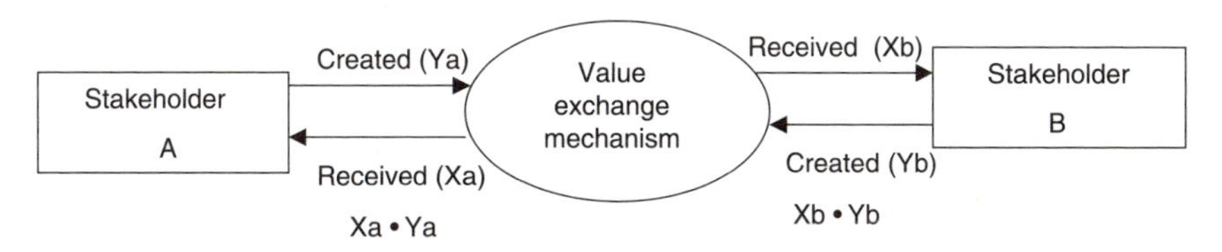

Figure 2.3 Balancing value exchange mechanism.

manageable, one can argue this will serve to provide a constructive stimulus in further value creation, based on the notion that manageable risk can create further opportunities for investment. However, if the imbalance or the difference is very large, this can lead to a dysfunctional WLC value exchange mechanism. A large difference could also be seen as an indicator of the risk of decay (or lack of common objectives and trust) in WLC stakeholders' relationships.

It is also evident from Figure 2.3 that it is impossible to maximise or optimise received and created value(s) independently. That is between the two value exchanging parties and within individual firms. According to Miller and Lewis (1991), avoidance of any domination by any WLC party's interests is essential to the attainments, convergence and maintenance of value exchange equilibrium or balance. It follows from this argument that to achieve an equitable value balance is to maximise the value(s) exchange through principles listed above. However, if all WLC participants use ethical practices for the benefits of the involved companies and society in general, an increase in value will be realised for all parties involved. Having said this, the concept portrayed in Figure 2.3 demonstrates that exchange values must be maintained in essential balance as a prerequisite to exchange. This should be achieved through continuous improvement, benchmarking of quantity and quality of values received by each WLC stakeholder. Such improvement will require long-term commitments from all WLC parties.

Further development

The nucleus of an equitable WLC value exchange process is a set of WLC relationships based on trust and common objectives. In this work, value exchange is viewed as a multifaceted dynamic process rather than a static mechanical process. This is essential for a better understanding of WLC interdependencies and complexity in WLC value exchange. This will allow WLC parties to understand all types of WLC value exchanges and avoid concentration only on the economical aspect of the exchange process.

The work presented in this chapter has added a detailed analysis and description of the WLC value exchange process to the prior research. It has further speculated about aspects that may lead to disequilibrium in WLC value exchange processes. The approach presented here emphasised that all WLC parties must seek to enhance the exchange value of their WLC operation based on equitable and balanced mechanisms.

The WLC value exchange framework presented here provided a means for managing WLC tangible and intangible benefits by shifting the focus away from traditional practices of costing and cash flow problems (i.e. economic value) to how WLC stakeholders' value can be maximised through collaborative effort. WLC parties can use the outcome of the WLC value exchange process in connection with value analysis to identify where opportunities exist to increase revenues, reduce costs, and increase capital, knowledge and physical asset utilisation for all WLC stakeholders. The value exchange process described in this chapter can also be used to measure both internal and external value exchange mechanisms in terms of the value(s) created and the value exchanged by all WLC participants. This is essential to identify equity and imbalances that exist in WLC relationships and their associated risk. Identifying risks in WLC value exchange is very important in directing WLC management efforts to restore and maintain equitable WLC value exchanges.

One implication arising from this theoretical work is that further research should be aimed at extending our understanding of WLC value types and the role of WLC stakeholders in creating value for themselves and for others in the WLC chain. An equally important aspect of WLC value exchange is improved understanding of the value drivers in the WLC process. Another potential research topic is a deeper investigation of the risks associated with value exchange. The value research agenda should also seek to align its objectives with the necessity to engage stakeholders in understanding that the built asset itself is not a sufficient measure of value. The research groups should look at innovative ways of encouraging those involved in preparing business cases for capital projects to shift the focus from demonstrating the value of 'cost' to how the asset contributes to a wider purpose, that is to say, complements a service or provision which adds value to business, society and environment.

Summary

Whole life cycle exchange is a relatively new concept in the management of performance. This study provides a theoretical background for WLC value exchange and proposes a rational ratio-based formula for measuring value exchange. However, WLC value exchange is a concept that requires much additional research to support the theoretical work proposed in this chapter.

Chapter 3

Whole life cycle value creation

Introduction

Built environment stakeholders have demonstrated considerable interest in methodologies for formulating built asset procurement strategies based upon WLC philosophies. The notion of 'WLC protocols' for developing the business case for a built asset requires a coordinated, synchronous approach in order to achieve higher levels of WLC services for stakeholders at lower total whole life cycle cost (WLCC) and is an inherently attractive proposition. This goal is usually accomplished through investing in building assets that produce net benefits that are readily calculable such as a positive cash flow that lies above the original WLC budget estimate (Boussabaine and Kirkham 2004). Logically, this condition is only satisfied when the discounted WLC benefits or cash flow exceed the cost of the capital and the WLC budget in addition to other intangible benefits. If these conditions are satisfied, then investors/stakeholders may believe they have added value to their respective organisations. However, this constitutes only one facet of value and ignores intangible value benefits that are the reason why project stakeholders should be rewarded for undertaking and delivering projects that return over and above that which is specified in the WLC budget, including the cost of capital, through the whole life span of assets, thereby creating value to the client organisation. Hence, the concept of maximising value for client investment through added value performance measures. Performance measures of WLC value creation will assist the construction industry in developing value strategies to the benefits of clients, end users and the built environment in general.

WLC value creation is crucial not only to enhance client and end user loyalty and return margins (value) for the WLC stakeholders, but also to ensure the very survival of the WLC stakeholders' organisation in an increasingly competitive built environment market. Hence, the challenge for the built environment lies in the need to create value throughout the WLC process – from the 'cradle to grave'. In order to serve the built environment stakeholder better and achieve the desired value in whole life cycle processes, traditional boundaries will have to be shifted and barriers between stakeholders broken down. This requires a different vision of whole life value and an understanding of how value creation changes with time. The previous chapter addressed some of these issues and presented an overview and definition for value alongside a model for WLC value exchange. The work also proposed a model for WLC value convergence. This chapter adds to the value exchange framework presented in Chapter 2 by introducing a conceptual model for WLC value creation and value maximisation followed by a discussion on risk variables that influence WLC value creation.

Defining whole life cycle value creation

The definitions of value(s) and value exchange were identified and described in a structured manner in Chapter 2. The WLC value creation issues are discussed in the rest of this chapter. Bowman and Ambrosini (2000) have combined extant bodies of theory into a coherent explanation of value creation and value capture. These authors argued that the inputs of tangible and intangible resources into the life cycle process are incapable of transforming themselves into value. It is only through exploitation that value is created (Peppard *et al.* 2000). Value is created by the action of projects stakeholders. The action may represent a construction process or provision of a service within the life cycle that generates value. The value of the action depends on the life cycle process in question, how it enters into the value exchange mechanism, as discussed in Chapter 2. The magnitude of value created is only determined at the point of exchange. The created value cannot always yield added exchange value to the value creating stakeholder. Value creation in financial terms is based on measures such as net present value, payback time or internal rate of return balanced against risk. This definition is very narrow and excludes all other intangible value.

For the purpose of this work, WLC value creation is defined as a continuously evolving process to maximise a stakeholder's value in conjunction with the asset value objectives through the life cycle process. Stakeholders' value is discussed in Chapter 2. Cha and O'Connor (2005) provided a comprehensive discussion and a modelling tool to assist in achieving asset or project value objectives.

The concept of whole life value creation having a time dimension has not been the subject of extensive investigation or debate within the built environment sector. Figure 3.1 shows how whole life value is created and exchanged over time. The philosophy of value portrayed in Figure 3.1 is that whole life value evolves over the life cycle of assets or services. The value creation and exchange model presented here must have clear measures or metrics for assessing value at each key milestone of asset or service life span. Note also that the values of stakeholders change as the asset progresses through its natural development process. The idea here is that the value placed on any building asset or service may increase or decrease and also the composition of the value exchange delivered by the asset/service may also change and emerge over the life span. For, example, if a building asset becomes obsolete for any reason, then a new use for

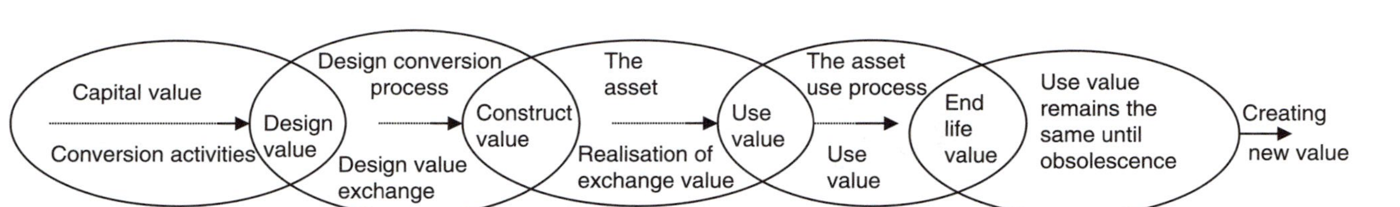

Figure 3.1 The whole life cycle value creation process.

the asset may be found and hence the emergence of a new value. The consequence is that the initial value and benefits that are perceived at an early stage of the asset lifespan may decrease in value and at the other extreme it is also possible that they may become irrelevant. It is also possible for an unexpected value to emerge for different stockholders. It is crucial for asset/service stakeholders to capitalise on the existing building assets/service by actively seeking to create new value(s) and benefits that outweigh the operational financial management (FM) and maintenance costs. If new value(s) are constantly created throughout the lifespan of building asset/service, then the service life of the asset will be extended and service providers will benefit from further business. This inevitably will lead to better use of resources and contribute to a sustainable built environment.

The question is, therefore, how can new value(s) be created? If we accept that building assets or WLC services can be modelled as a portfolio of investment through its life span, then at each stage the portfolio stakeholders must seek to unlock the value in building asset investment through several whole life mechanisms, including lean management, process regeneration, etc. New value can also be created by changing the perception of what constitutes value (Bannister 2001). The following sections are an attempt to conceptualise some of these issues.

Cycle of value creation at the design stage

Designers create services that are valuable to their clients. Design has been considered a major influencing factor in the final building asset form, functionality, WLC budget and creation of value (Boussabaine and Kirkham 2004). More explicitly, the arguments are that the major WLC value creation decisions are made in the very early phases of the design process. To generate the expected value asset, design should simultaneously satisfy various WLC aspirations, including functionality, cost, reliability, constructability, maintainability, marketability, profitability, flexibility of use, reusability, etc. The main challenge for design stakeholders is to support these multiple WLC aspirations to create value for their clients while producing value for themselves (and potentially other stakeholders). As illustrated in Figure 3.2, the definition of value creation during the design process might be divided into four specific value creation processes. In the first instance, customers; individuals or organisations recognise that there is a need to invest in a building asset to add benefits and improve their quality of life or add value to their businesses. These needs must be fulfilled by addressing and finding optimum solutions to solve the problems that have led to the emergence of these needs in the first place. This is the antecedent of any value creation and exchange. Second, professional designers use their intellectual capital (value) and practices in combination with financial capital and land to find a design solution to the problem (needs) previously identified. Design stakeholders create value by using their professional services and skills in support of their customers. Cha and O'Connor (2005) developed a tool that

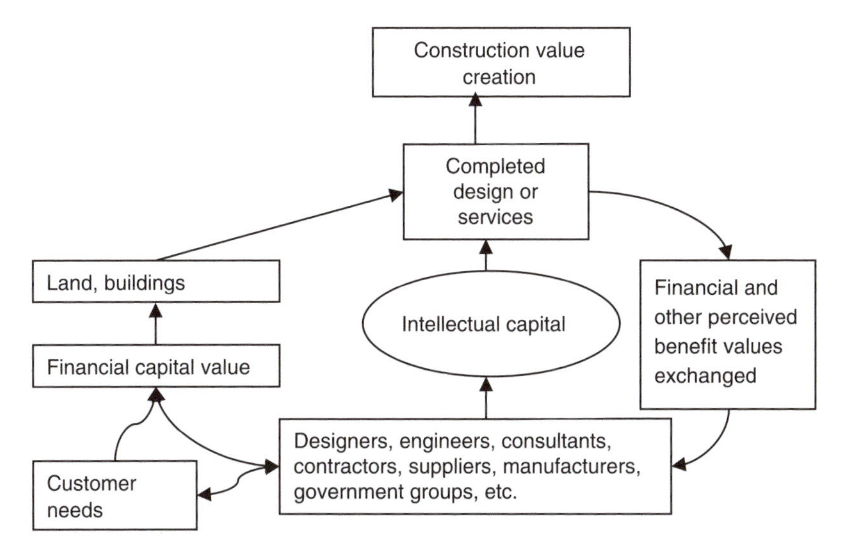

Figure 3.2 The cycle of value creation at the design stage.

assists in achieving project value objectives. The value creation competencies or indicators at the design stage are defined by CABE (2002). These include functional, technical, financial, operational, and social factors.

The design team must adopt value analysis mechanisms to identify design improvement (creating value) in terms of quality, performance and WLC cost reduction opportunities. The design stage does not represent the end of the WLC value creation's process. This stage will lead to the second hierarchical level of WLC value creation.

The completed design or design services can be considered as a value exchange milestone. It is possible that only part of the design process is complete and the value creation may be developed alongside the production process. Whatever the process, the result of this stage must be satisfactory and profitable to all clients and service providers, i.e., shared added value to all businesses involved.

Cycle of value at the production stage

The exploration of the concept of value creation at the production stage is illustrated in Figure 3.3. First, several production stakeholders consider entering into an exchange with each other. Each party will have their own perceived list of benefits that they would like from the value exchange mechanism, to enhance their possibility of creating value for themselves and ultimately for their client and society in general.

The tangible inputs (design capital value, financial capital value, intellectual capital value and other resources) into the construction process are converted by construction organisations to create new use value(s) from the acquired

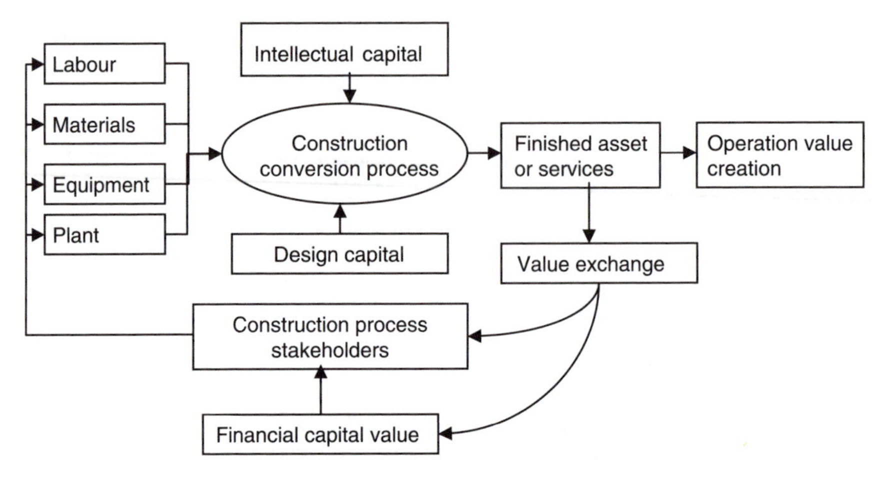

Figure 3.3 The cycle of value creation at the construction stage.

resources as shown in Figure 3.3. Hence, value is created by the actions of construction production participants, who combine to transform the design capital values that clients have acquired into operational value or use value. It is important here to stress that construction stakeholders must realise added exchange value, that is, the value of the constructed asset is greater than the costs of all the resources expended so far on the asset.

Cycle of value at the operation stage

The extent of value created by acquiring a new building asset can only be determined when the asset is put into use or exchanged (sold). At this stage, asset value will be assessed by potential users/occupiers as to how it achieves their perceived benefits and adds value to their businesses or quality of life. As illustrated in Figure 3.4, operational value is created through the effective and efficient use of business operation, engineering systems, building systems, and FM systems. Here, as can be seen from Figure 3.4, the sources of financial capital value and human capital support the operation of the building asset by investing in the operation cycle to maximise value returns. It is imperative to state here that if the value capital generated by business operation just covers the operational needs of the business and building asset, the conversion/exchange process does not create value to the occupier/user/owner of the asset. If the operating cycle shown in Figure 3.4 does not operate efficiently, the chances of attractive receivables (added value) providing an increase in value to all operation stakeholders will be minimum. The figure also depicts the addition of returning gross values from operational benefits, etc. But, in some cases, the building asset will be operating with a negative margin. In such cases, instead of creating tangible financial value, all supporting operation activities will use capital value.

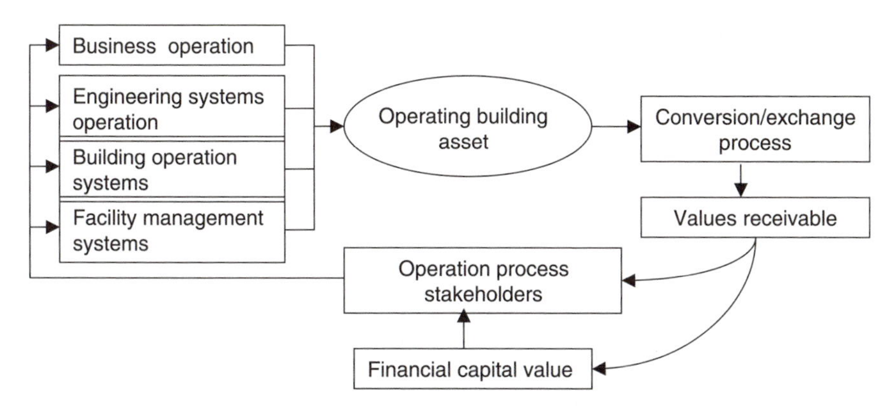

Figure 3.4 The cycle of value creation at the operation stage.

Cycle of value at the end life stage

Building assets are usually procured and/or operated over a long period of time. Over this period, equipment, building components and assets used in the operation of the building facility become obsolete for many reasons (see Boussabaine and Kirkham 2004). Figure 3.5 provides a conceptual model of the overall cycle of value creation when a building asset reaches its end economic life. Here, there are several options available for end life stakeholders to capture value. The most important principle here is to try to maximise value/return from these options.

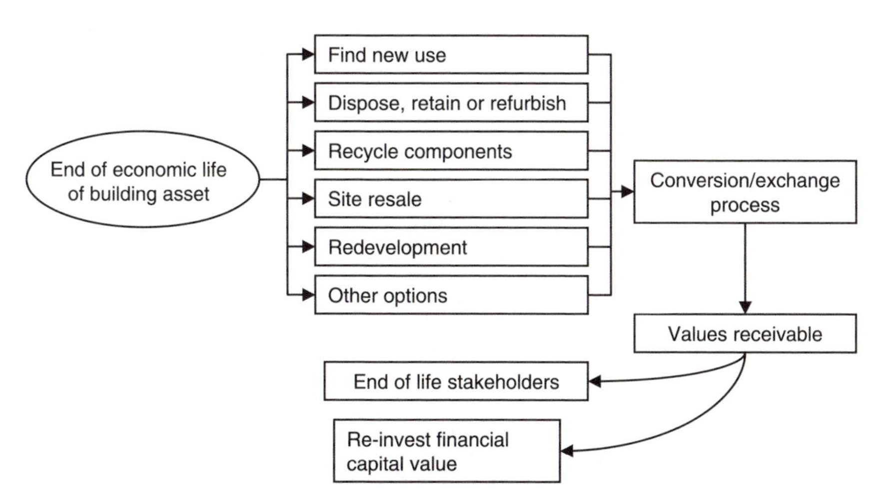

Figure 3.5 The cycle of value creation at the end life stage.

Maximising WLC value

The maximisation of WLC value should be the ultimate objective of all WLC stakeholders. To maximise WLC value, the WLC participating firms need to effectively and efficiently select and manage human, tangible, and financial whole life resources, processes and assets as well as services to maximise the aggregate additions to WLC stakeholder value.

All WLC participating organisations should accept that their primary task is to maximise value returns to themselves as well as to others in the WLC chain. Hence, the concept that all WLC participants should align their activities to the pursuit of this ultimate fundamental objective. If this conjecture is accepted, then the key role of WLC stakeholders is to develop ways to maximise value for their clients and simultaneously for themselves.

In Chapter 2 we proposed equations for measuring value created and value revived in the WLC chain. We advocated that the concepts used in the equation should be used to maximise the value of the product/service, conceived as a ratio between the end users' satisfaction and the costs for the value creator. Thus, value is created by satisfying the end user as well as minimising costs and prices used in the creation and exchange of value. For value creators, value might be maximised if the firm profits and workload increases, plus other intangible benefits, whereas for value receivers, value is maximised through increasing satisfaction in product or service at a minimum/optimum price. In these equations, dissatisfaction of any WLC firm can lead to a decline in the value relationship and trigger a decrease in value creation.

The attributes that contribute to WLC value maximisation are portrayed in Figure 3.6. As illustrated in here, WLC maximisation can be conceptualised as two interactive sets of attributes relating to value creation and value received.

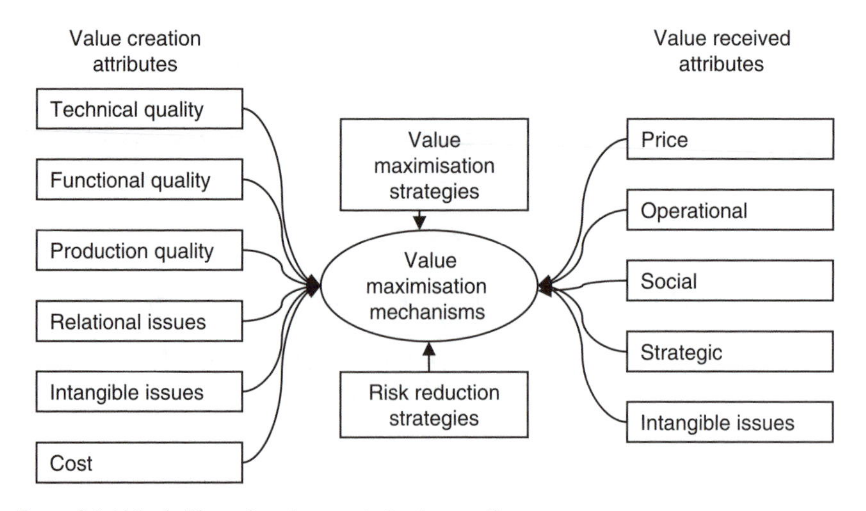

Figure 3.6 Whole life cycle value maximisation attributes.

Figure 3.6 indicates that value creation is best maximised through value creation and value received attributes. WLC value creation characteristics can be maximised through the technical and functional construction quality of the product/service being created. Also, relational attributes such as partnership, trust, common objective, etc. and capital cost spent on creating value are imperative in value optimisation. Other intangible factors such as corporate image and credibility are equally critical.

Value received (in use) can also be maximised through factors such as price, social, operational, strategic, image and all other perceived intangible benefits that lead to the satisfaction of WLC stakeholders.

Both value creation and value received attributes must be maximised simultaneously in conjunction with WLC maximisation strategies and value risk reduction variables.

In using value as a measure of performance in the whole life cycle built environment, stakeholders can maximise whole life value for their clients by:

- increasing whole life operation profits without increasing whole life budget;
- reducing whole life cycle budget;
- increasing whole life budget, if increased capital charges are more than offset by increased profits;
- achieving optimum combination of whole life cycle cost and quality to satisfy value receiver requirements;
- increasing WLC alliances and relationships with clients, strategic partners, suppliers, investors, built environment communities and society in general;
- improving WLC value knowledge and experience across all the participating firms;
- increasing emphasis on WLC long-term value objectives without sacrificing current value and performance of each WLC organisation;
- increasing service life and value in use of the asset.

Risks to value creation in WLC

Long-term investments are subject to various sources of whole life risk, influencing the potential outcome of investment or services. Risks to value creation are viewed as a multidimensional concept entailing multiple types of risks including financial, performance, physical, psychological, time and social risk (Snoj 2004). Financial risks can be categorised as risks that a WLC stakeholder is losing capital value, because the service or product exchanged does not satisfy the perceived needs. Psychological risks are related to the choice of service or product selection and procurement. If the wrong product or service is chosen, capital value may be lost and might have other negative effects on the whole WLC chain. Functional risks can be attributed to the state of the product or services provided. If the service or product failed to fulfil its functional requirements as expected by the receiving WLC firm, then all or most of the invested value will be lost. Physical risks are related to the building asset condition over the lifespan of the capital value investment. Social risks are attributed to the obsolescence of WLC service and products and hence the negative effect on the value. Time risk relates to the time spent in providing a service or developing a product that does not perform according to a particular WLC organisation's viewpoint. Further discussion on risks that might affect WLC value creation can be found in Boussabaine and Kirkham (2004).

Relating risk to WLC value, it is inevitable that many factors can affect whether or not a particular WLC value can be realised, for example, as a built asset. These risks may lie in any phase of the life of the asset from inception to disposal, or indeed span several phases. One can argue that the rule of thumb is, that the greater the perceived risk of any built asset, the less the value (particularly the economic value). Conversely, it is also held that the lower the perceived WLC risk, the higher the WLC value (see Figure 3.7 and Figure 3.8).

It is also important to point out that the mechanism by which WLC risk is allocated and transferred will alter the value WLC stakeholders assign to building assets or WLC conversion processes and services. It is very risky to invest capital value (both financial and human) in building assets during design and production operations. This riskiness is due to the fact that it is very difficult to capture value from the incomplete building asset. That is why at every point of value exchange shown in Figure 3.7, if exchange does not occur, value will be destroyed. However, once the asset is complete, the risk to value creation is lower. However, the risk to value returns from the operation stage still remains high, as shown in Figure 3.7.

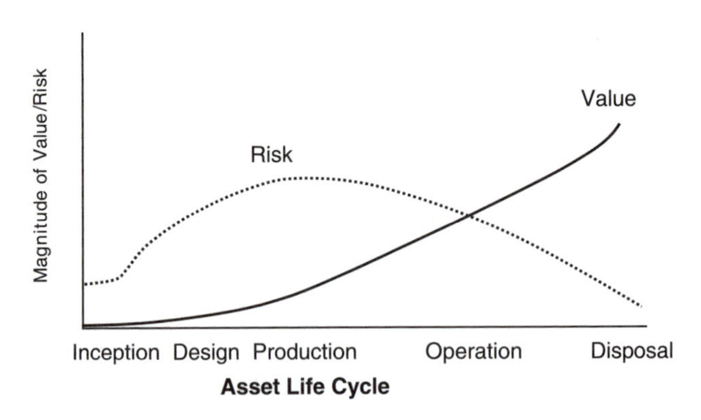

Figure 3.7 Conceptual risk and value changes during the life cycle.

Source: reproduced with the kind permission of J. Lewis of the University of Liverpool, School of Architecture.

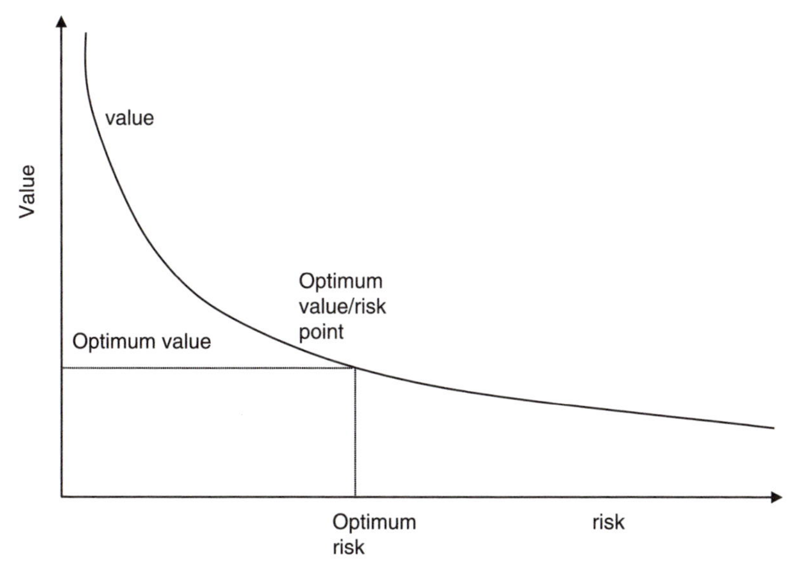

Figure 3.8 Relationship between value and risk value.

Summary

In this chapter the we have proposed the notion of WLC value creation. We have embedded our discussion in the perspective of the whole life cycle paradigm and have taken the view that effective WLC value creation should be viewed by WLC stakeholders as a strategic asset. This chapter has also contributed to the WLC value agenda through an examination of strategies for WLC value maximisation and value-influencing risk factors. The work presented here has also advocated the concept that the key to WLC value creation resides in life cycle stakeholders' perception of value, rather than in procurement and operation costs. WLC stakeholders, when engaging in WLC value creation, should balance the built asset's net benefits (tangible and intangible) with its whole life costs. The greater the ratio of net benefits to WLC, the higher the value.

The research value agenda and debate within the built environment sector should concentrate on value targets that need to be developed within the life cycle agenda. In doing so, participating organisations should pay attention to the fact that inflation of WLC value targets may lead to penalisation of some stakeholders purely because they were expected to deliver a higher value to client. The question one needs to ask is at what level values should be set. Perhaps the answer to this question lies in benchmarking current WLC values against what is being achieved within a particular sector of the construction industry. The objective here is to enhance delivered WLC value to all stakeholders. Before we embark on value benchmarking, we need to find out how much WLC value is actually delivered at project and industry-wide levels. The research agenda also needs to concentrate on finding mechanisms for delivering a framework for sustainable WLC value creation. All these topics require further research.

Part II

Procuring building assets under the PFI/PPP model

Chapter 4

Procuring projects under the PFI/PPP system

Introduction

The contract aspects of a PFI/PPP deal are probably the most difficult to plan, design and manage because of the complexity that arises from the interaction between procurer, lenders, investors, contractors, suppliers, insurers and those responsible for the operation of the asset. An essential step in the cost planning of PPP/PFI projects is planning a good contract design and efficient contracting procedures. It is safe to say that these two aspects are the foundation on which the whole paradigm of PPP/PFI is based. Well-designed and well-managed contracts are the vehicle by which risks are controlled and allocated between the public and private sectors. Incentives and penalties for management of contractual services are also directly linked to contract design. The notion of value for money also depends to a large extent on the precise terms and conditions of the contract, how the contract is planned, managed and awarded and how it will be implemented in practice over the project's life. Well-planned contracting procedures are also an essential ingredient in the reduction in time and costs of the procurement process. This chapter introduces the characteristics of PFI/PPP contracts.

Procurement options

The search for novel mechanisms to deliver infrastructure projects effectively has led to the development of new contract methods. A range of standard contract forms has evolved from traditional contracts and from private sector efficient practices that aim to maximise benefits for stakeholders through risk transfer and financial incentives for the service providers. The paradigm here is to transfer as much risk as possible to the private sector. Several conceptual models have been used to procure assets under these initiatives. These concepts are variously denoted as BOOT (build, own, operate and transfer), BOT (build, own and transfer), BRT or BLT (build/lease and transfer), DBO (design, build and operate) and DBFO (design, build, finance and operate). All these contract systems need to be structured and managed so as to fulfil the objective of delivering value for money to the clients. The following subsection will discuss the characteristics of some of the widely used models of contract.

Traditional contractual model

A traditional contract has the following features:

- under this mode of contract, the procurer assumes and manages all risk;
- public sector cost of capital is generally lower than the private finance;
- experience shows that projects delivered under this scheme tend to be over budget;
- behind schedule and lacking in innovative thinking, particularly in terms of whole life cycle issues;
- incurs financial burden (adds to public debt).

Figure 4.1 shows the contractual relationship between project stakeholders. It shows that the public sector enters into several contractual relationships in order to produce a particular asset or service. All these contracts are for the short term and lack long-term commitment from the involved parties. Also, in most cases, performance risks relating to cost and time overrunning are not contracted out to the private sector. Operational service standards and performance measurement risk will reside with public sector bodies.

Design, build and maintain (DBM)

In DBM contracts, the private sector provides assets and ongoing maintenance services but the public sector pays for the asset on completion and for the maintenance services when provided over a medium term contract. A DBM contract will have the following characteristics:

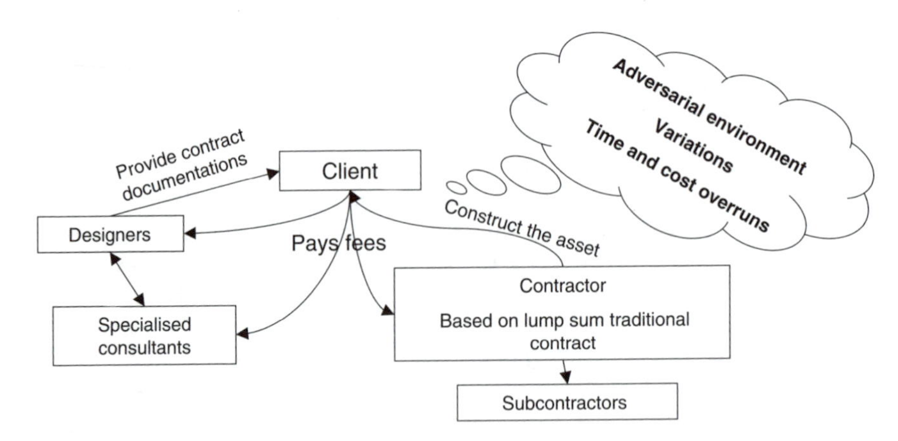

Figure 4.1 Traditional contractual relationship.

- The private sector will maintain the asset and also be in charge of capital repair and replacement.
- The public sector will arrange their own financing strategy based on a bond issue or other appropriate methods.
- The project may be run by a publicly owned non-share capital corporation.
- The corporation may enter into a concession agreement with the appropriate public authority for a prescribed concession period. This is usually done with the purpose of matching the term of funding.
- Separation of the terms of the DBM agreement from the terms of debt.

Figure 4.2 illustrates the contractual relationships under this mode of procurement. It is obvious here that there is less complexity of financial transaction which can lead to a reduction in the cost of capital procurement as well as the cost of the procurement process. It also leads to the loss of private sector skills and competition due to lower contracted services. Cost of capital could be higher compared to DBFO. Although the integration of contracts and medium-term maintenance risk transfer may result in savings in capital and operational costs, achieving a high level of value for money through risk transfer to the private sector is very difficult without contracting out other services like operations.

Design, build, finance and operate contractual model

The concept of the DBFO contractual model is based on the idea that instead of the public sector developing a capital asset and providing a public service, the private sector creates the asset through a single stand-alone special purpose vehicle (SPV) company (see Chapter 5) to deliver public services, in return for payment linked to the performance and levels of services provided. As well as designing, building and providing services, the private sector will also arrange debt financing from commercial lenders for a high share of the cost of the asset

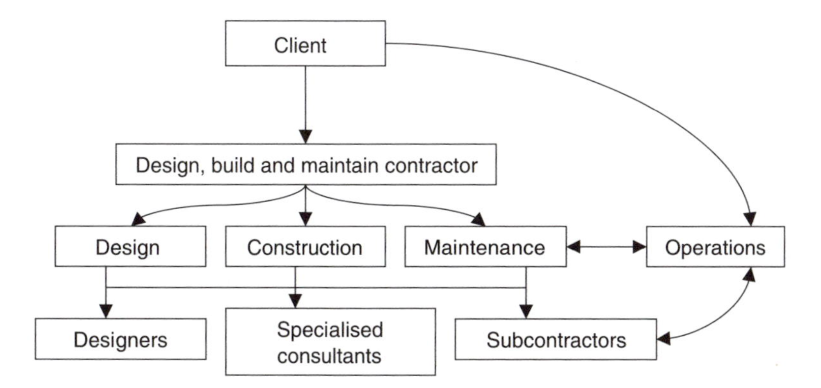

Figure 4.2 Design, build and maintain procurement model.

and equity for the balance of the funding requirements. The central characteristics of a DBFO contractual system are that the private sector supplier assumes the provision of public services including the design and construction of any building asset and the associated operational management of the facility and its accompanying services based on an output specification from the public sector. The private sector is also responsible for the overall management of financing the entire development, including the development and construction where high capital costs have to be incurred. A DBFO contract will have the following features:

- The provider gets paid on completion.
- The public sector pays a capital charge over the contract life which is used to repay the lenders and to pay the equity.
- It is a long-term contract where the private sector enters into a concession agreement with the public authority for 25 years or more.
- Asset ownership by the private sector during the concession agreement.
- Usually the concession agreement includes design and construction of the asset, whole life maintenance for plant and fabric, capital repair and replacement, and provision of soft facility management services.
- The concession agreement includes details on provision on payment, service standards and performance measurement, and provides an objective means of varying payment, depending on performance.
- The private sector bears the risks of arranging funding.
- The private sector is incentivised to perform well and deliver value for money through equity investment. This will also provide security for investors.
- Ownership reverts back to the public sector at the end of the concession with the possibility of extending the concession agreement.
- Assets must be maintained to a specified standard for the whole life cycle of the concession and must be returned to the public sector in a predetermined condition.

Figure 4.3 demonstrates the contractual relationships in the DBFO contract model. This model is created based on a single point of responsibility of an SPV. This structure will allow the private stakeholders greater freedom to determine the means of delivering the specified public sector requirements. Hence, the central core of this mode of procurement is that there must be one public sector and one private sector entity party to the contract. Usually, however, there may be more than one private sector entity involved in delivering the facility and associated services. As shown in Figure 4.3, executive responsibilities are typically delivered through sub-contracts to the principal private sector party (SPV) to the DBFO contract (Yescombe 2002). It is understood that by using such contractual arrangements, assets and associated services will be delivered on time, on budget and to a higher standard. Maximum opportunity for

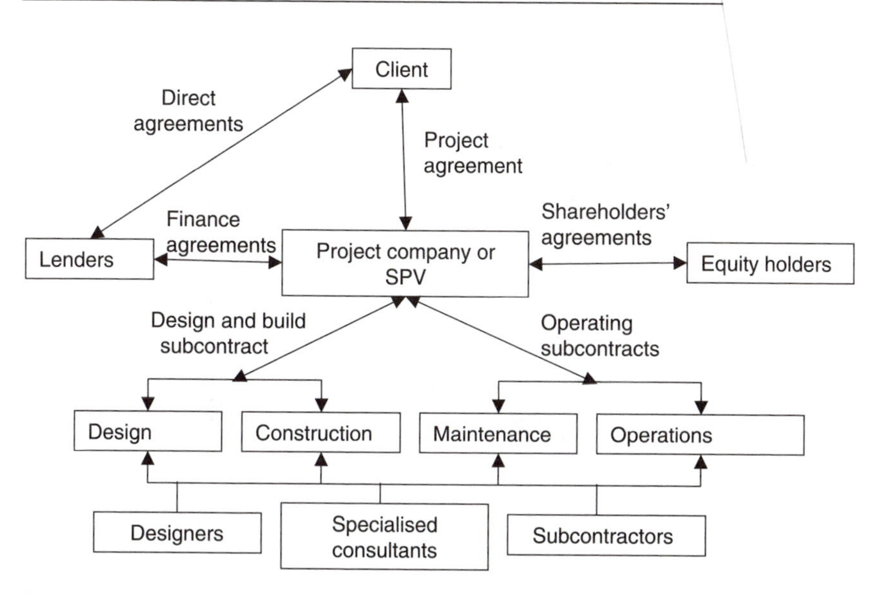

Figure 4.3 Design, build, finance and operate contractual model.

innovation and risk transfer to the private sector will also be enhanced. The DBFO contract model reduces and converts capital expenditure in the form of payment for service delivery; as a consequence, the taxpayers will get better value for money and cost-effective provision of public services, at least in theory! In return, the public sector will benefit from increased and additional business opportunities and from securing long-term revenue with potential high profits if the whole life risk is managed appropriately. However, one of the drawbacks of this procurement method is that the terms of contract are determined in terms of financial arrangements and not asset considerations which may have adverse consequences and additional risks when the asset reverts back to the public sector.

Further details on financing mechanisms, SPV and the role of each stakeholder are presented in Chapter 5.

The PFI/PPP procurement process

The procurement planning process for both public and private sectors are illustrated in Figures 4.4 and 4.5 respectively. The following sections will explain briefly the planning requirements for both sectors and identify the key milestone stages of the process and some of the actions that need to be taken to deliver a successful procurement process.

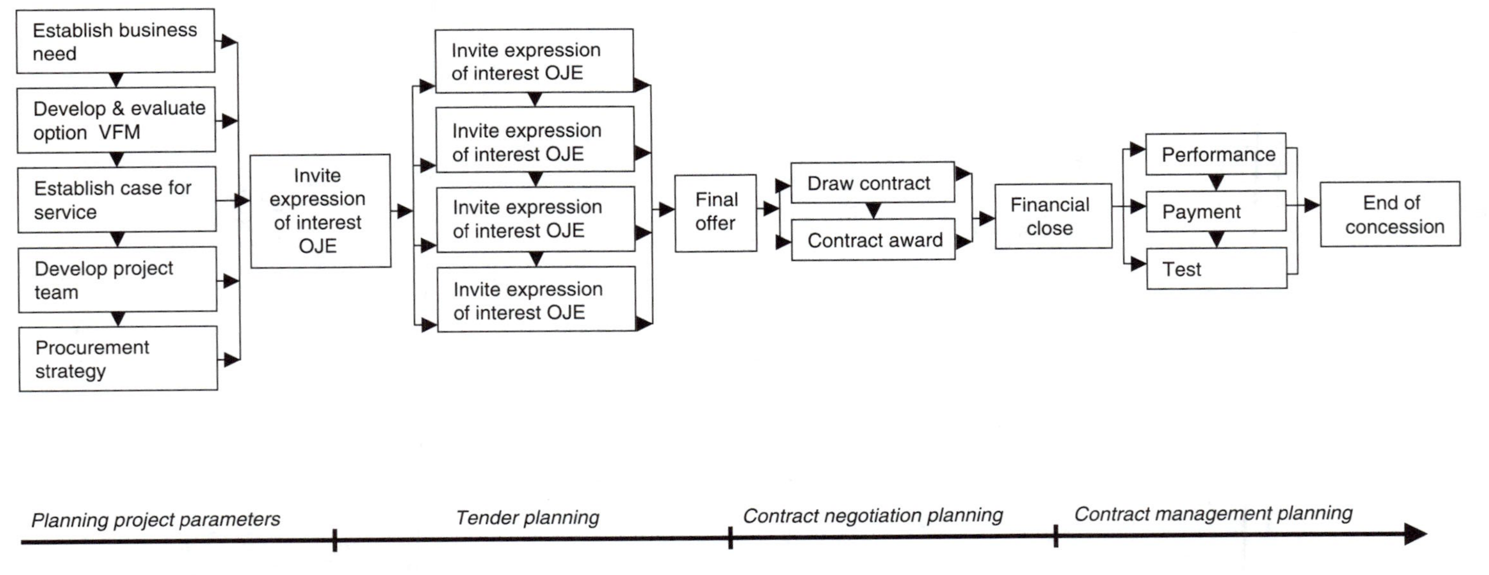

Figure 4.4 The public sector PFI/PPP procurement map.

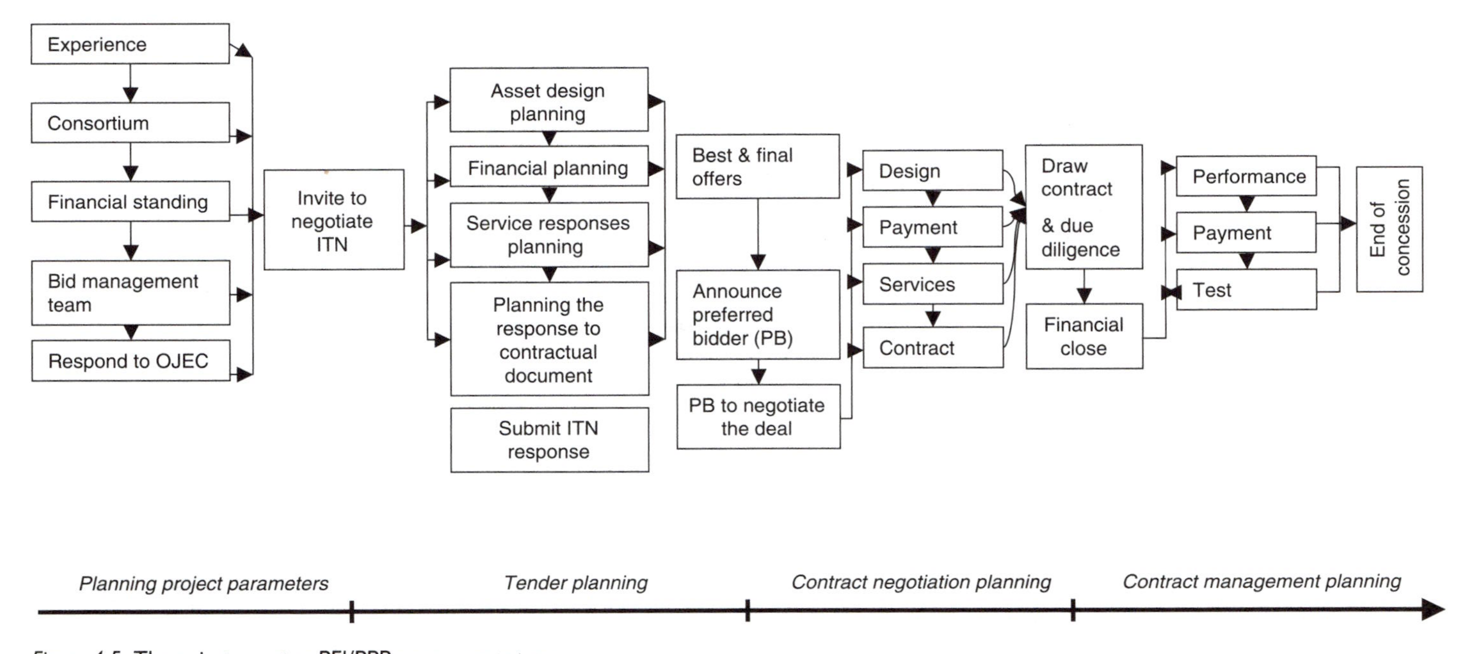

Figure 4.5 The private sector PFI/PPP procurement map.

Public sector

As illustrated in Figure 4.4, the procurement planning process for the public sector can be divided into four key stages.

Identification of project parameters

The planning process for project parameters identification shown in Figure 4.4 takes the form of a hierarchy of steps that the procurer, or in this case the public sector, should follow to get a good deal from PFI/PPP procurement systems. The planning process starts from the general declarations of establishing the business needs and making the case for services to be procured. In addition to a detailed description of the services to be provided, many procurers incorporate as a schedule to the contract a description of their needs and the service output requirements that need to be met. The details, specification and context of each of the tasks shown in Figure 4.4 will increase and expand in scope as the procurement process progresses. Establishing business needs requires the procurer to state clearly what he is looking for from the intended project and how he expects this outcome to be delivered (NAO 1999). The procurer needs to set clear objectives and provide answers for all the whats and hows. In establishing service needs, the procurer must do the following:

- Identify the volume and quality of the service required and the full costs (capital and operational) of their provision over the whole life cycle of the service or asset.
- State clearly the project deliverables.
- Focus only on what outcomes/outputs will improve public services and which provide a good deal. Under the PFI/PPP paradigm, the inputs are the responsibility of the private sector.
- Recognise all regulatory constraints on how a particular service is to be managed and delivered.
- Develop a set of meaningful benchmarks/criteria against which project benefits are constantly assessed.

It is common practice for most of the public sector departments to produce an outline business case before the announcement of tenders. The outline deals with objectives, desirable outputs and benefits of proposed PFI projects. This should include an assessment of whether the project is viable and whether a PFI solution would offer the best value for money. The aim of the outline business case is to do the following:

- Set out clearly the objectives of the project, as described above.
- Carry out a form of appraisal to demonstrate that the PFI/PPP approach is the best option for delivering value for money (see Chapters 2 and 3). This

must include an affordability test assessment of whether the financial benefits of a PFI project are reflected in increased value for money over the whole life of the project and not simply in savings on short-term capital expenditure.

- Produce service performance requirements.
- Produce a thorough plan and timetable for the procurement process.
- Establish evaluation criteria for assessment of tenders received.
- Set out clearly all necessary financial commitment, if any, to pay for the project once delivered, and produce scope of satisfying the contract structure test.

The next step in the hierarchy process is the determination of the best procurement strategy. Forms of possible procurement strategies are discussed above. The aim here is that the public procurer should assess what is best for their needs before going into the market. This will help to reduce the cost and time scale of the procurement process. In selecting the best strategy, the public sector procurer needs to consider:

- risk transfer and allocation to deliver better value of money;
- innovative ways of delivering the required services.

The next step is the assembly of an effective qualified team to oversee and manage the procurement process. This is essential on all projects but particularly imperative to the successful completion and capture of VFM in PFI/PPP projects. The team should be composed of multi-skilled, experienced negotiators who have participated and have the skills to deal with the private sector. The team should include experts with knowledge of how to deal with EU and UK procurement regulatory systems. It is very common for the procurer to appoint external advisers with previous experience in PFI/PPP work to assist house teams in providing good legal and financial advice. The team should be responsible for:

- setting timetables for all stages of the procurement process;
- monitoring progress;
- negotiating the deal;
- drawing up the contract;
- resolving problems as they arise;
- investigating the market for the services in question with a view to developing a procurement strategy of how to present the project to the market;
- considering issues that are likely to arise in negotiating the deal.

Usually the procurement team is led by a senior member who represents their interests and acts as the project's focal point for the day-to-day management of the project. The procurement team is usually split between core service (e.g. education, transport and health), technical, and legal and financial, with sub-groups formed to address specialist issues. The outcome of the planning project parameters stage will be the issuing of an invitation for expression of interest. The tendering/procurement may commence with a notice in the *Official Journal of the European Community* (OJEC).

Planning tenders

The procurement process begins with the project notice in the OJEC. Meanwhile the procurer will be active in producing a marketing brief, describing the project and the form of procurement in more detail, and a preliminary questionnaire. These documents are sent out to all those companies who express their interest in the project. The procurer's tendering team evaluates the returned questionnaires based on a predetermined set of criteria. The procurer usually pre-selects a number of bidders for further interviews. The number of pre-selected bidders is largely dependent on the number of firm submissions received. Based on the interview results, the number of bidders will be reduced to around three bidders; however, the quality of prequalification submissions and the market sector will determine the number, as there may be a limited number of players able to deliver the services. These bidders will be issued with the instructions to negotiate (ITN) documentations. The ITN have to be returned by a pre-determined date. The public procurer may conduct further interviews with the bidders if necessary. The invitation to tender should be thoroughly checked, cross-checked and completed. This will reduce issuing amendments during the tender period as well as the need to clarify a tender during appraisal.

The ITN documentation is exhaustive and usually includes the following important parts:

- *Instructions to negotiate*: includes information about the procurement process, compulsory items that have to be considered or included in the bid and the evaluation criteria for selecting the preferred bidder.
- *Building design output specification*: this document lays down the design requirements for building assets and associated accommodation concerning the operational management and its relevant policies, and the operational and capacity requirements.
- *Building service output specification*: this comprises all the performance requirements and quality standards for the estates and management service; catering services; caretaking, security and safety services; and the equipment provision and maintenance services.
- *Contractual framework*: the basis on which the contract will be awarded, such as the standard contract model.

After this, the procurer will revise the section criteria, usually developed at an early stage, if necessary. It is common practice that the selection criteria for the preferred bidder are clustered into legal, financial and technical aspects. These criteria are assigned separate weights. However, in some cases they are equally weighted. Thus, the purchaser uses bids to compare the offerings of each bidder against the affordability limits in the outline business case and test assumptions about value for money. The evaluation process is undertaken in stages to quickly identify key issues and provide an overview of the attractiveness of each bid to the procuring department. It is normal for the procurer to select a preferred bidder (PB) and a reserve bidder. In some cases, the reserve bidder will be involved during the negotiation process. One of the bidders will be awarded preferred bidder status. This will lead to the contract negotiation and ultimately financial close.

Contract negotiation planning

The aim of this process is to reach a financial closure as soon as possible based on a sound legal framework. It is well recognised that the total cost of tendering for a PFI project to all potential suppliers is over 2 per cent of expected total costs while for traditional procurement the total costs account for just less than 1 per cent. This is attributed to the fact that the time taken between offering public services projects to the private sector and the final signing of the deal is unacceptably high. The average time taken to complete PFI deals is over one year, and in some cases may extend to over three years. This suggests that the public procurer spends a considerable time negotiating with preferred bidders, which has the knock-on effect of an increase in total expected costs to the purchaser. To address these concerns, the UK Government has for some years been involved in developing the standardisation of the tendering process. The aim of standardising the tendering process and contact documents is to allow future PFI contracts across different public services to follow a consistent approach by incorporating standard conditions into all contracts. The UK HM Treasury published the first edition of *Standardisation of PFI Contracts* in 1999. This was revised and republished in 2002 by the Office of Government Commerce in 2002. The third edition was released in 2004 and is based on the experience of the public and private sectors since 2002, and seeks to incorporate a number of improvements that have been identified over this period (HM Treasury 2004a). Among the aims of the contract document is to reduce the time and cost of negotiation by enabling all the parties concerned to agree to a range of areas that can follow a standard approach without extended negotiations. According to the National Audit Office (1999a), to reduce the length of time taken to close PFI deals, the procurer should do the following:

- demonstrate a clear purpose and a strong vision of the desired outcomes from the scheme;

- establish simple output specification and eliminate or minimise changes to specification;
- gain early commitment to the scheme from key stakeholders;
- develop a project management structure that allows for an appropriate level of delegation to key officers and is integrated with existing decision-making processes (as discussed above).
- establish a robust project plan with project milestones and monitor progress against the plan on a regular basis;
- establish and agree the key contractual terms, including payment mechanisms and risk transfers, prior to issuing the invitation to negotiate, in order to force bidders to indicate their position early on in the negotiation process.

The following issues may cause a delay in the negotiation process:

- A stakeholder may join the negotiations too late and make decisions that have to be revised in order to have the deal financed.
- Lack of commitment from stakeholders.
- Lack of motivation of teams that drive the negotiation process.

Contract management planning

Managing long-term contracts over a 20- or 30-year term is one the most challenging aspects of PFI/PPP procurement and needs careful short- and long-term planning. Effective contract planning and management must be fundamental to the success of any public–private partnerships. It is important here to stress that the foundations for contract management are laid during the early stages of procurement planning processes. The issues that need to be addressed and planned for are associated with operational performance of suppliers, payment methods testing and monitoring of the quality of services. The OGC (2005) has listed the factors that should contribute to successful contact planning and management as follows (some of these issues will discussed in detail in Chapters 6 to 10):

- accurate development and assessment of project parameters (as described above);
- effective tendering procedures and robust criteria for selecting providers, based on clear output-based specification rather then traditional input requirements;
- creating and planning for an effective long-term relationship between the procurer and suppliers;
- clear roles and responsibilities and dispute resolution mechanisms are required for the essential fulfilment of this aspiration;
- effective planning for risk reduction and control;

- development of effective methods for monitoring the contract's performance and effectiveness, including management of non-performance;
- establishment of effective methods cost control and control of payments.

It must be stressed that failure to adapt to the above long-term procurement strategies can result in contractual disputes and, according to NAO (1999a). could lead to:

- problems in contract interpretation;
- poor performance/quality service/output specification;
- failure to agree a process for new additional services;
- delays/late delivery and missed key deadlines;
- changes in requirements;
- disagreement over responsibilities;
- poor communication.

Private sector

Figure 4.5 illustrates the procurement planning process for the private sector. The process can be divided into four key stages.

Bidding strategy planning

The bidding process is based on strict rules and procedures from host countries and sometimes, in the case of Europe, from the European procurement regulations as well. The process is complex, competitive, lengthy and costly. The decisions of a particular supplier as to whether or not they should bid for PFI/PPP projects is a key strategic undertaking with a huge capital expenditure commitment over a long period of time. In making such decisions, providers should fully realise the opportunities and limitations provided by these types of contracts. Opportunities offered by such procurement systems include availability of large investment by the public sector, high-value long-term contracts with a good return and the possibility of generating further capital through the securitisation of income, lower risk due to fact that the volume risk is retained by the purchaser and a standard contractual form that provides reasonable security against disputes. Private suppliers must also recognise that there are a number of potential threats such as high bidding costs with no guarantee of success, accepting long-term risks based on the fact that all contract provisions are based on fixed price, and long procurement processes. With the sheer volume of opportunity coupled with the complexity of the process, the providers need to have effective and efficient plans and mechanisms in place to produce a winning bid. A typical procurement process for the providers is shown in Figure 4.5. The first step in the planning process is to assemble a well-qualified and experienced team to respond to the OJEC advertisement. The aim here is to build a winning consortium that can show extensive experience in the type and scope of services that they are trying to bid for, a track record in delivering successfully similar or major projects and an indication of commitment to the bidding process from the consortium members. Normally a bidding consortium will consist of:

- a construction team;
- a facility management team;
- funding teams, from both senior debt and equity funding organisations;
- advisers for legal, technical and financial aspects of the bid.

The first task of the consortium is the completion and submission of the pre-qualification questionnaire with associated documentation such as companies' accounts. In the majority of cases the pre-qualification questionnaire consists of two parts. The first part assesses the technical, financial and economic strengths of companies expressing an interest in bidding for the provision of the specified services and accommodation. The second part requests information on approaches to service policy and project-related subjects. To pre-qualify for the next stage of the process, procurers need to plan and concentrate on the following aspects:

- relevant experience and details of each supplier, their roles and key individuals;
- PFI experience and quality accreditation of each provider;
- details of the providers' shareholders and equity provision (or potential shareholders/equity provision);
- turnover, capacity and technical experience of the providers' designers and technical advisers;
- financial capacity and relevant technical experience of the providers' facilities management services operator(s) (this can be audited accounts over a specified number of years);
- financial capacity and relevant technical experience of the providers' construction contractors and mechanical and electrical contractors;
- capacity and technical experience of the providers' construction project manager;
- capacity and technical experience of the providers' legal adviser;
- capacity and technical experience of the providers' financial adviser, and the providers' fund-raising capability;
- asset planning, design and construction;
- partnership working arrangements;
- change in management procedures;
- statement on the understanding of services provided and how the providers will assist the purchaser in delivering improvements in its services.

Planning the negotiation – ITN stage

Based on the assessment of the pre-qualification questionnaire, a number (usually up to three) of providers will be asked to submit an ITN. The client is looking to the bidders to demonstrate that they understand the scope that is required. The ITN will be issued by the purchasers and consists of the following elements.

Executive summary

Bidders are required to provide an executive summary of the bid submission. The document may not include any information detailing or summarising the price of the bid. Providers may be required to provide a separate financial summary sheet, providing basic information on the unitary charge for each facility. A concise summary should also include design and operational proposals, consortium members, response to legal aspects, and management and organisation of the bid team. The document might be limited in format and number of pages; hence, the summary must be upbeat and include all the key aspects of the project including an explanation of the added value that will be generated by the proposal to all stakeholders (see Chapters 2 and 3).

Project management

Bidders are required to provide responses to the criteria such as a strategic approach to set out how the purchaser's strategic objectives will be addressed through the provider's proposal for the project. For example:

- The bidder must indicate what he can bring to the project by way of skills or innovative solutions to meet the success criteria set by the client.
- Integration of design and facility management to demonstrate how consistency will be assured between the design and VFM solutions (bidders may be asked to demonstrate this concept through real examples of PFI schemes).
- Management of stakeholders to set out the roles and responsibilities of managers within each service provided (including both the construction of facilities and facilities management) and how these will be integrated into one coherent whole (the bidder must make clear the selection process of sub-contractors).
- Proposals for their replacement, market testing and performance measurement and integration within the output specification.
- Collaborative working with the purchaser to set out an overall strategy and mission statement as a long-term provider of the specified service (details of long-term partnerships including management structures prior to and after financial close).

- Management of the programme to financial close to provide a detailed programme that identifies all the key tasks required to achieve financial close, their timescales and inter-relationships (this must include the management process that bidders will use to ensure the programme is maintained).

Legal response

Providers need to plan their responses to the draft project agreement and associated schedules. It is important for the provider to understand that usually all negotiation in relation to the project agreement with the purchaser will focus almost entirely on project-specific design, commissioning, and technical and service issues. This will primarily be required in order to develop and complete all the requisite schedules to the project agreement. Bidders should also consider project-specific issues such as risk allocation, insurance, planning issues, etc. Providers must indicate what project-specific amendments they may wish to raise together with their justification and possible rewording of a particular clause.

Financial response

Providers are required to respond and submit to purchasers' information requested in the ITN finance section along with the financial model for the reference bid and any variant bids submitted. In most cases, the financial submission requirements are set out in two key parts: *the financial structure and the deliverability of the funding package*.

The procurer needs to ensure that providers have access to the finance necessary to support the bid submitted. Hence, providers should include full details of their proposed funding structure, setting out clearly the institutions involved, the type of facility to be provided, the amount of funds available and the terms on which these funds are being offered. Bidders may be required to submit a copy of the formal documents confirming the funding terms, including all fees, offered to the borrower from the funder. This applies to the providers of all classes of finance including equity, subordinated debt and mezzanine debt, in addition to senior debt. The suppliers must also demonstrate in detail their strategy for ensuring the procurer gets a value-for-money funding solution, through funding competitions, benchmarking or some other mechanism. Firm financial commitment and the strength of support from the funders are essential at this stage. Bidders must make sure that the financial backers support, accept and agree with the contents of the provider's financial model, payment mechanism and project agreement. Providers also need to plan and provide support from their financial adviser to certify that the proposals are realistic and achievable on the basis of the project agreement, payment mechanism and the other bid submissions, assuming there is no adverse material change in market conditions.

Costed bid proposals

It is a requirement for procurers to compare submitted bids against each other and against the PSC. Hence, bidders are required to provide financial projections for the full duration of the contract period (from financial close to end of concession or a specified number of years) in the form of a dynamic financial model. The financial models are based on some consistent key assumptions and provide certain fundamental outputs. Bidders therefore must comply with the format and assumptions set out by the procurer in the ITN documents. It is expected that the financial models are compatible with the requirements of funders (including periodicity and calculation of ratios) and should be capable of being used to run sensitivities in the key areas of risk that the funders and rating agencies are likely to focus on. Providers in general must plan their responses for the following financial aspects of the bid (these aspects will be discussed in some detail in Chapter 5):

- details of financing structure;
- information on equity, subordinated debt and mezzanine debt;
- hedging strategies;
- funding flexibility;
- parent company guarantee;
- deliverability of funding;
- financial model requirements/assumptions to include model format; price base; economic assumptions; cost allocation assumptions; taxation and accounts assumptions; funding costs; model output specification; financial performance; model databook and sensitivity analysis.

Bidders should include the following information to the procurer:

- *Design and construction*
  The bidders are required to plan and provide sufficient design and construction details to enable the procurer to assess and evaluate the proposals and the providers' approach to managing the obligations and responsibilities as set out within the ITN documentation. The providers must plan and deliver the following:
  - Town and Country planning issues, comprising: conformity with policies; design; siting and layout; parking and traffic circulation; landscaping; highways and access, etc;
  - design of the building content of the proposals, comprising: internal specification; flexibility; external specification; functional/departmental areas; gross floor area; whole life approach, e.g. life expectancy, maintenance, etc;
  - design of the engineering services content of the proposals, comprising: design deliverables; energy policy; environmental standards and

specifications; systems flexibility; life expectancy, maintenance, and commissioning;
  - cost assessment issues, comprising: construction costs; equipment costs; contingencies or risk assessments; non-works costs; fees allowances; whole life costs, etc;
  - capital and development costs cash flow.

- *Facilities management and human resources*
  The key difference between traditional procurement and PFI/PPP is in the provision of ongoing service requirements. Hence, it is paramount that bidders must provide comprehensive service statements on how they are planning to maintain the asset and meet the output specification over the concession period. In their submission, the providers must include comprehensive details on the following aspects:

  - maintenance strategies, both reactive and planned;
  - compliance with the requirements of the procurer's service output specifications;
  - service delivery proposals;
  - proposals for ensuring appropriate quality of service delivery;
  - Human Resources – staff employment/transfers and transition management (Retention of Employment); ongoing management, etc;
  - service specification – supply, installation, commissioning, maintenance and replacement;
  - VFM investment plan, cost plan and financial model;
  - customer and procurer liaison arrangements and procedures;
  - security and complaint handling procedures.

- *Variants bids*
  It is good practice if providers can supply a schedule detailing all areas where their reference bid does not comply with the procurer's reference bid requirements, statutory regulations or the advisery requirements set out in the ITN documentation. A clear statement of departures usually accompanies each variant bid from the bidder.

- *Additional information*
  Bidders might be asked to provide a media pack and information for inclusion in any public consultation document that the purchaser may want to carry out. Providers may also be required to provide descriptive copy and graphics of the design proposals that they are willing to place in the public domain.

  Providers in some cases are expected to provide a range of materials to support their bid submission and assist the purchaser in the evaluation process. The additional supporting information might include:

  - models of the sites and buildings;

- artists' impressions of the landscaping, main entrance, key public spaces and building façades;
- visual 3D simulations of key areas, including public spaces;
- samples/examples of external finishes;
- any other relevant materials that demonstrate value-for-money aspects such as sustainability issues.

Clients may ask bidders to provide method statements because of the nature of the specifications used in PFIs that are written in output terms. The client is looking to the bidders to demonstrate that they understand the scope that is required. The client also expects bidders to satisfy requirements in terms of service availability, quality of service delivery and responsiveness. Through bidders' responses, the client has to be given enough information to evaluate the strength and capability of each bidder. Although method statements are not the only tool the client relies on (there is also reliance on record of accomplishment and financial stability, for example), the method statement is used to differentiate the management and operational capabilities of bidders.

Planning the negotiation – PB stage

The contractor and his lenders are faced with a considerable workload between preferred bidder selection and the contract award stages; a realistic and achievable timetable for this period should be agreed upon at the earliest opportunity. The BAFO stage is used to negotiate and refine the key commercial and financial terms. Formalisation and standardisation of documentations of the bidding aspects discussed above will allow bidders and clients to limit negotiations to the key commercial terms, not technical ones.

It is essential at this stage that stakeholders produce a detailed negotiation brief and key points that are needed to be discussed with relevant stakeholders. This will assist bidding teams to negotiate within agreed parameters and swiftly agree on solutions. The issues that should be included in the negotiation briefs might include risk allocation, variation, payment mechanisms, etc. Inconsequential issues should be left until essential matters are negotiated.

To speed up the process, it would be beneficial for all parties, if possible, to combine the commercial and financial contract closings. This enables the commercial and financial agreements to be looked at as a package in negotiations, with the result that compromises could be reached with minimal impact on the overall contract. If this is not possible, usually the PB stage is followed by the financial and commercial close of the project. It is a requirement here that the funders appoint financial experts, usually accountants, and technical experts to carry out due diligence on the bid. Financial experts will audit the financial model for consistency, accuracy, sensitivity, and so on. Technical experts usually carry out audits on the construction programme and maintenance proposals. If funders are satisfied with the outcome of the auditing process, the concerned parties then will be in the position of closing the deal. It is normal practice at this point of close that the unitary charge is fixed by reference to base rates.

Contract management planning

The process of contact planning management is discussed in detail in Chapter 10.

Contract structure

PFI/PPP project contracts are complex and subject to numerous subcontracts within the overall framework of the project agreement. These contracts, from the point of view of the purchaser, must achieve flexibility and place emphasis on output specifications. The project contracts provide the legal framework under which the providers (SPV) build, operate and provide services to the procurer. There are several project contracts relating to project finance, construction, etc. It should be stressed that the following sections are not about legal issues that might be addressed by these types of contracts, but concentrate on the issues that are included in them for planning and cost planning purposes. For a detailed debate on various legal and technical aspects of project financing and other contractual issues, readers should consult Fight (2005) and Yescombe (2002).

The project agreement

The project agreement is the most important PFI/PPP contract document because it provides a legal framework under which the project SPV obtains its revenues. According to Yescombe (2002), there are two important models for a project agreement:

1 *An Offtake contract* is the principal contract between the provider and the procurer under which the Offtaker/buyer/procurer buys the product produced by provider. The Offtaker is obliged to take delivery of not less than a specified minimum quantity of the product over a specified period. The obligation here is on the provider to deliver to the Offtaker the agreed or specified quantity of the Offtake on the date and at the stated price agreed at the time of entering the contract. Such agreements provide project revenue stream for the seller and a secure supply of the required product for the Offtaker. These contracts are risk-sharing instruments between producers and buyers. There are several forms of Offtake contracts. Since this book is mainly for PFI/PPP projects that are procured under concession agreements, specialist readers can find more details in the work of Fight (2005) and Yescombe (2002).

2 *Concession agreements* are also referred to as service agreements or project agreements. This is the primary contract between the procurer and the concession company and forms the contractual basis from which the other contracts are developed. It entitles the project company to build, finance and operate the facility and provides a service (not a product) to public purchasers or directly to the general public. In return, the procurer will pay a unitary charge to the project company based on service availability and quality of services provided. Penalties are payable if the services are not to the required specification. Hence, the definitions of both availability and the quality of the service to be provided are very important aspects of this contractual agreement (see Chapter 5). Going back to the first principle of the PFI/PPP paradigm, to add value through risk transfer to private sector, the project agreement leaves certain risks with the project company as long as these do not reduce the contract revenue to such an extent that debt service and a reasonable return on private investments are undermined or put at risk. Typical risks that normally reside with the providers include: project time and cost overruns, project/service availability and operating costs (see Chapters 12 and 13 for more details).

Terms of a project agreement

Typically, terms of a project agreement will include the following (for more details, see, for example, the standard PFI contracts by the UK HM Treasury).

Preliminaries

This section discusses issues related to the terms of a project agreement, interpretations and definitions of contract terms, commencement and duration, and the operations and issues related to execution and delivery of project documents.

Contract preliminaries lay down the rules on how contract terms have to be interpreted within a particular project agreement. They are project-dependent. Definitions and interpretations of the project agreement terms are usually included in schedules. The commencement of the project agreement normally coincides with the date of signing the project agreement. The provider's obligation to carry out the project operations terminates automatically on the expiry of the project duration. The purchaser can terminate this obligation at any time, if the provider is unable to fulfil his obligation under the project agreement. The most significant aspect in this section is the 'term' of the project agreement. This is the duration of the PFI/PPP project's contract, that is, the period until the final repayment date of the debt. The starting point (date) of the contract operation is important because several computational issues may depend on this date. For example, the mechanics of finalising interest rates for funding arrangements, financial model assumptions, term of the debt, service life of components, useful life of the asset, residual value of the asset, etc. The UK standard contract form assumes that the contract signature and financial close will be simultaneous. If financial close and the contract signature are not simultaneous, owing to the existence of specific project circumstances, the provider and purchaser will need to decide how risks relating to interest rate fluctuations between contract signature and financial close are to be allocated. Another complication which stems from this is redefinition of the project 'term'. Any variation on the interest rates could have impact on the pricing provisions for the procurer and return on investment for funders. Here, the parties may need to consider if the project 'term' will commence at financial close and not the date of the signature of the agreement.

Other issues that are covered in this section include control over changes to the funding agreements after the contract is signed. The obligation and responsibility for any changes to funding agreements and refinancing are dealt with here. It is expected that the procurer will usually obtain a share in refinancing gains in accordance with the criteria specified in the contract schedules (see Chapter 11).

Provision for the project company to carry out its obligations and the right to perform its duties at its own cost and risk without recourse to providers are also included. However, a project agreement may provide clauses, in exceptional circumstances, where providers may have a right of recourse to purchasers. Procurers' undertakings to perform their operational and financial obligations, plus mechanisms for co-operation between the contract parties, are normally incorporated into this part of the agreement.

General provisions

This sub-division may embrace matters such as warranties, indemnities and liability, limits on liability, representatives, liaison and disaster plans.

The UK standard form of a project agreement does not envisage that third party revenue generation activities will form part of PFI/PPP projects. Hence, if revenue generation is anticipated, then the standard form of agreement needs to be modified as appropriate to cater for any additional activities to be carried out as part of a project.

Since the SPV must be solely created to contract and operate PFI/PPP projects, the project company must provide warrants that it has not traded at any time since its incorporation as a company. This will also extend to the accuracy and authenticity of any information provided, usually in a separate contract schedule, about the project company.

The SPV is responsible at all times for indemnifying the procurer against all direct losses, as specified in the project agreement. The losses are project-specific. It is the responsibility of the SPV to manage the risks covered by the indemnity by putting appropriate insurance cover in place (which it is required by law to do), which contains satisfactory non-vitiation provisions. The procurer is equally responsible at all times for indemnifying the provider for any losses specified under the agreement. The purchaser and the SPV should insure against these risks as a matter of good contract management, and the procurer must review the relevant subcontracts to ensure that they contain mirror provisions excluding the right to claim indirect losses.

The procedures for conduct of claims, mitigating circumstances, and guidelines on limits on liability are also described under this section of the agreement.

Both contracting parties are expected to appoint or name their representatives to deal and pursue the issues included in the agreement. A liaison committee from both parties is to be put in place to oversee all issues relating to implementation and operation of the agreement as well as providing recommendations on strategic changes and variations to the existing agreement to accommodate changes and improve efficient implementation of the agreement.

Land issues

This section deals with the nature of land interests, the site and consents and planning approval (PA). The main issues here are the selection of site and access rights. The standard form of PA assumes the procurer selects the site. The provider will require rights to access the facilities during the operational period. The access right is usually restricted to the performance of the project operations and it is normally based on a non-exclusive basis. The access right starts from the date of issuing the certificate of commencement and includes occupation of the site and gaining access to and from the site. It is very important to point out here that standard PA does not place any restriction on the provider in relation to the use of the site, provided that the provider complies with his obligations under the PA. Hence, the extent of specific site access restrictions are largely negotiated between the contracting parties on a case-by-case basis. The issue that needs to be addressed here is who pays for the sewerage, rates and utilities during this period (Chapter 10 provides further details on these issues). Issues relating to the grant of leases and compliance with the title deeds are also dealt with under this section. Under standard PA forms, the condition of the site is solely the responsibility of the provider. This is based on the condition that the site is a greenfield site and that the provider is able to make all necessary investigations. If this is not the case, then areas of the site, which the provider is unable to investigate or identify, ought to be carved out and clearly identified in the PA. It is normal practice in such circumstances for the procurer to bear any additional costs arising out of these unforeseen conditions from the identified areas. Under PA, it is the responsibility of the provider to obtain detailed consents and the planning permission required for the performance of the project operations.

Design and construction

This section sets out the responsibility for the design, construction and commissioning process, right of access, programme and dates for completion, independent tester, provision of equipment, pre-completion commissioning and completion, post-completion commissioning and quality assurance.

It is the responsibility of the provider to carry out the design and construction of the projects to the satisfaction and requirements of the procurer based on the provider proposals and with accordance with the terms of the PA. It is common practice that the procurer usually reviews the final version of the design prior to contract signature. This is essential so that procurers can reassure themselves that design proposals satisfy their requirements in terms of functionality, durability, life service of components, etc. If these proposals do not fulfil the procurer's specified requirements, then it is the responsibility of the provider to amend the proposals at his own expense (see Chapter 10). Under PA, the provider is obliged to allow unrestricted access at all reasonable times to a procurer's representative to inspect and view the works at the site. The representative also has the right to attend construction progress meetings. The procurer is also responsible for complying with all relevant safety procedures and legislation. This part of PA also sets out clearly the programme and dates for completion. It is the author's understanding that the PA standard forms do not impose pre-quantified damages for late completion. The same also applies for liability for damages due to late completion. These issues are usually addressed on a project case-by-case basis. However, provision for notification of early completion has been included in PA standard forms. If an early completion occurs or is envisaged to occur, then this will have a knock-on effect on aspects such as commissioning programmes, phasing, equipment procurement, payment and other issues relating to the procurer's early occupation of the facility.

The terms for the independent tester are laid down in this section of PA. It is common practice for the independent tester to be appointed jointly by the contracting parties; some procurers may stipulate that the independent testers must act as experts with bidding decisions on both parties (see Chapter 10).

Clauses on provision of equipment are project-specific and deal with who installs, who is responsible for what type of equipment, parties' liabilities for installation of the SPV or its contractors, the impact of delay by relevant parties on equipment installation and all issues related to ongoing equipment requirements.

Regarding pre-completion commissioning and completion, the UK standard form of PA envisages the following sequence of events (further details are also included in Chapter 10):

1 The Final Commissioning Programme (FCP) agreed. The FCP will set out all requirements and obligations in relation to the development, nature,

principles and performance of the completion tests to be performed to enable certification of completion to take place.

2 Provider commissioning needed in order to achieve practical completion.
3 Procurer to carry out its commissioning activities scheduled to take place prior to completion (if any).
4 Provider to carry out completion tests in accordance with FCP.
5 Visual inspection by procurer and independent tester.
6 Certificate of 'Practical Completion' and problem list issued.
7 Commencement of appropriate percentage of the unitary charge.
8 Provider completion commissioning activities in accordance with FCP.
9 Procurer post-completion commissioning in accordance with FCP.

It is widespread practice for the contacting parties to usually consider which commissioning activities have to occur before and which after completion.

In post-completion commissioning, the parties need to consider what commissioning/services start-up activities will be carried out by the provider and what commissioning activities the procurer will carry out during this period. The payment mechanism and performance monitoring systems are structured in such a way as to reflect the intention of the parties in relation to services starting up during this period.

Provision for quality assurance is provided in the form of a framework in the standard PA. The stipulation for quality management systems must be consistent with BS EN ISO 9001 or 9002 or any equivalent standard. Normally, the quality systems include a design quality plan, a construction quality plan, and a services quality plan for each service provided. The roles and responsibilities of the quality manager are also included in this section.

Services

This consists of aspects such as specification of services, maintenance, monitoring of performance, site security and personnel issues, stocks, consumables, materials and equipment and value or market testing.

The clause clearly sets the responsibility and obligation of the provider to provide contract services in accordance with the PA and the service method statements (see Chapter 11). The contracting parties agree to a schedule of programmed maintenance. The need for and the terms of a five-year maintenance plan in some cases are required. Unless project specifics determine otherwise, there should normally be no deductions during periods when planned preventative maintenance is taking place in accordance with the agreed schedule of programmed maintenance. Normally the risk in relation to maintenance and replacement lies with the provider (see Chapter 13).

Monitoring performance clauses set out rules on deficiency points, warning notices and their consequences. It is expected that detailed performance monitoring systems will be provided as a separate schedule to the PA. The PA standard form includes some of the procedures for the periodic testing of the price for services. However, some procurers opt for other alternative methods of value testing, such as benchmarking of services (see Chapter 11).

Payment and financial matters

This section provides reference for payment, insurance, provider insurances, custody of the financial model, information and audit access, changes in law and variation of procedures.

Most procurers use the principle 'only pay for what you receive'. Based on this principle, a sequence of events is followed. Practical completion of the works triggers commencement of payment of the unitary charge. In cases where a project is completed in phases, the payment mechanism is structured so that payment of a relevant percentage of the unitary charge commences on practical completion, with build-up of the payment being linked to completion of each phase of post-completion service commissioning. This means that an increase in payments is triggered by completion of each phase until full payment is reached.

Issues relating to invoicing and payment arrangements, manner of payment, disputes, late payments, amounts overpaid or wrongfully paid by a party are dealt with in this section (Chapter 5 has further details on payment modalities).

The provider is required to procure insurances at his own cost. These are taken out prior to the commencement of the works and are maintained for the periods specified in the insurance requirements schedule as part of the PA. In return, the procurer is required to specify in detail the key insurances the provider is required to procure. Providers on a commercial basis insure liabilities relating to facilities constructed and/or provided under a PFI/PPP schemes. Clients are, in some cases, permitted to be named as an insured party on policies which providers are required to take out and the premium for such policies are by and large included in the service payments charged to the procurer by the provider. Provisions for uninsurable risks and risks that may become uninsurable risks are addressed here. When uninsurable risk occurs, the procurer may pay to the provider an amount equal to the insurance proceeds. However, the procurer would not pay the deductible (or in excess of any limit) to which the insurance would have been subject. In extreme circumstances, or *force majeure*, the PA may be terminated on a non-default termination basis.

The custody of the financial model is provided in accordance with the provisions of the 'custody agreement' between the contracting parties. Both parties have the right to inspect and audit the financial model. Amendments to the model should be in consistent PA.

Changes in law and variations

In case of any changes in law, it is the obligation of the provider to ensure that the project operations are performed in accordance with the terms of PA and compliance with law. However, in the PA standard form, the financial consequences of such changes in law rest with the procurer (including in relation to the cost of obtaining any necessary consent); the risks of obtaining the consents remain with the providers. This clause makes provision for the adjustments to the service payments to compensate for any increase or decrease in the net cost to the provider of performing the project operations. Variations are dealt with according to specific variation procedures described in the PA schedule (Chapters 10 and 11 have further details on cost variation at both construction and operation stages).

Force majeure

This section consists of matters relating to delay events, relief events and *force majeure*. Delay events are related to aspects that may have an adverse effect on the ability of the provider to complete the facilities by the completion dates and may includes issues such as procurer works variation initiated by a procurer works variation enquiry, relevant change in law, *force majeure*, etc. Dispute resolution procedures included in the PA as schedules are used to decide how compensation will be paid. Relief events are related to incidences such as fire, explosions, lightning, storms, accidental loss or damage to the works, official or unofficial strikes, etc. Usually only approved specified events are listed in PA. Statements that refer to any event outside the control of the parties should be a relief event and are not acceptable in a PA. This section provides provision for mitigation in case a party is affected by relief events. It is common practice that the payment and deduction mechanisms provide procedures on how service payments are reduced during a relief event to reflect poor or non-performance in relation to specified performance criteria. In general, financial deductions are applied during a relief event (this is usually subject to clauses provided in a PA). The reason for this is that in PFI schemes the financial risks of such an event lie with the providers.

Force majeure events are habitually associated with aspects of war, civil war, armed conflict, chemical or biological warfare, nuclear contamination, earthquakes, etc. If an event of *force majeure* occurs, the PA includes provisions for compensating the provider for losses. This provision is based on the general principle that services should be paid for to the extent they are received (i.e., the deduction mechanisms should allow for no payments to be made to the extent that no services are delivered). This stipulation includes clauses on how payments will be reduced during an event of *force majeure* and what payments the provider will continue to be entitled to.

Termination of the project agreement

This part includes aspects relating to provider events of default, procurer events of default, non-default termination, effect of termination, compensation on termination, and handback procedure. Under the PA standard form, the provider events of default are associated with insolvency; long stop (provider unable to complete the works within a period of 12 months after the completion date; default (breach of obligations by the provider under the PA); health and safety; change in control; assignment; poor performance; payment and notification. In case of provider default, the procurer has several options to deal with the circumstances including termination of the entire agreement with immediate effect; serving notice on the provider to remedy the default and to put forward a programme of action to put the project back on track; termination of a third service provider due to poor performance. However, there may be restrictions imposed on the procurer. For example, the procurer should note that any rights to terminate would be subject to the terms of the funders' 'direct agreement'. The defaulting party reimburses all costs incurred by the procurer in exercising any of his rights. Procurer events of default are linked to the breach of his obligations that materially and adversely affect the ability of providers to perform material obligations under the PA for a continuous specific period of time but normally not less than 30 working days; and the failure of the provider to pay any sum or sums due to the provider. The provider has at his disposal several options to deal with events including suspension of performance by him of his obligations under the PA and serving a notice on the procurer for termination of the PA with immediate effect.

Non-default termination deals with circumstances such as *force majeure* (see above), voluntary termination and expiry of the PA duration. PA termination is carried out in accordance with the clauses relating to continued effect – no waiver, continued performance, transfer to procurer of assets, contracts, etc., transitional arrangements, and continuing obligations.

Compensation on termination clauses sets out the rules for payment according to the default types discussed above. The rules for compensation on termination are usually set out in a separate PA schedule. It is common for compensation payments to be determined by reference to the return the shareholders/junior debt holders in the SPV would have received but for the termination. Payment of compensation to providers is subject to the procurer's rights under the PA.

The construction agreement

This agreement is between the SPV and its construction subcontractor. The difference between the concession agreement and the construction agreement is that the former is based on output specification, whereas the latter may well be more input-based, depending on the specifics of the project and VFM considerations.

It is the sole responsibility of the SPV to ensure that construction is completed on time and to budget. The SPV (not the purchaser) bears the cost of construction and the burden of maintenance and replacement costs over the life of the contract. The contractor, in turn, has the right to contract parts of the work to subcontractors. This might be, for example, design companies, specialist contractors or suppliers of key components. The evaluation and choice of these subcontractors are primarily the contractor's responsibility, but the SPV, financiers and the procurer will all have some input into the process. Under certain conditions, the contractor will have to seek approval for nominated subcontractors.

In this type of contract, the contractor takes total responsibility for design and execution of the PFI/PPP project. The contractor will carry out all the engineering, procurement and construction, providing a fully adequate facility, ready for operation. Construction will be procured under a fixed price turnkey design and build contract (a derivative of design and build contracts), with such construction obligations as exist under the concession agreement passed down to the contractor under the construction contract. The pricing of turnkey contracts is based on lump sum, cost plus and measured unit price. In PFI/PPP projects, the lump sum fixed price, turnkey contract, preferably with a specified completion date, is usually the preferred contract form of the SPV and project funders.

It is important to state here that the word 'turnkey' itself has no specific legal meaning. There will be liquidated damages for delay, based on the SPV's resultant loss of income.

Since the construction agreement in PFI/PPP projects is based on a design and build paradigm, most of the traditional contract forms in the UK such as JCT, ICE, NEC, etc. might be used for such a purpose. According to Capper (2004), a choice between these contract forms may be made based on the following:

- programme management;
- management of change arising during the works;
- commercial certainty;
- management effectiveness;
- legal consequences.

It is important to point out that these contract forms must be used in conjunction with the PA described above. This has the effect that the obligations provided in the PA must be translated on a word-for-word basis into the construction subcontracts. This process may lead to substantial amendments to the contract standard forms. Hence, this has led to the development of bespoke contract documents on a project-by-project basis.

The most widely used standard form of contract in PFI/PPP infrastructure projects is the EPC (engineer procure construct). The contractor is responsible for the design, procurement and construction. Yescombe (2002) states that the most important key elements of EPC standard contract forms are as follows:

- contract scope;
- commencement of the works;
- the project company's responsibilities and risks;
- contract price, payments and variations;
- construction supervision;
- definition of completion;
- *force majeure*;
- liquidated damages;
- suspension and termination;
- security;
- dispute resolution.

According to European International Contractors (2003), the aspects of EPC that most commonly give rise to disputes relate to the following:

- inadequate definition of the scope of work;
- disruption of the design and construction process;
- claims and dispute resolution.

Cost planning for a fixed lump sum design and construction contract requires a well-developed preliminary design and specification and clear and unambiguous performance criteria. The problem here is that the contract scope and pricing are defined by an incomplete design brief. Hence, any error or deviation from the initial assumptions will have a knock-on effect on procurement and contraction costs. Hence, it is to the benefit of both contracting parties to carry out the design in sufficient detail in comparison to the price of the works. The question to be asked here is whether it is possible to produce a detailed design of complex projects during the tendering and negotiation period! The answer may be no! This could have a direct effect on the way the works are priced. The contractor may inflate their prices to cover any unforeseen risks or liabilities.

The facilities management contracts

The facilities management contracts may be dealt with under one contract with a single contractor or separated into several subcontracts, if this is appropriate to the type of facilities to be provided. Facilities management contractors are often subsidiaries to the parent shareholders' companies. They are employed on similar terms to the main contract. The private sector bears most of the risk of providing VFM services. The SPV typically will enter into a long-term contract for the day-to-day operation and maintenance of the project facilities with a company that has the technical and financial expertise to operate the project in accordance with the cost and services specified in the PA. The facilities management contracts are not necessarily of the same duration as the concession agreement. It is important to point out here that the nature and extent of the service vary depending on the type of project. These contracts contain obligations to perform the SPV's operational obligations to the required specification, and to make good on any performance-related penalties. The operation contract may take one of the following forms (Tanega and Sharma 2000; Yescombe 2002):

1 *Fixed price arrangement*: All operational risks (with some exceptions such as changes in law service specifications, *force majeure*, etc.) are borne by the facilities contractors. This is an excellent arrangement for the SPV and may motivate the operator to optimise their delivery costs (adding value) without comprising quality of service provided.
2 *Incentive/penalty arrangement*: The contractor repayment is based on the availability of service and is usually computed based on an agreed formula for the unity charge (see Chapter 5). The facility management contractor may be entitled to bonus payments for extraordinary project performance and be required to pay liquidated damages for project performance below specified levels.
3 *Cost plus arrangement*: This model is based on a fixed fee plus the incurred operation costs. The level of fee is usually tied to performance.

The basic elements of a facilities management contract are:

1 Scope of contract and definitions
2 Services requirements
3 Fees and payment arrangements
4 Commercial duration of the contract
5 Obligations and responsibilities of the contracting parties
6 Incentives and penalties
7 Arbitration
8 Termination.

Schedules to the contract normally include documents supporting the implementation and conclusion of the contract (see Nanayakkara 2002, for further details). The exact content of these contract elements is highly dependent on which form of agreement is used and to a certain extent on the operator's role in the SPV and, more importantly, these are drafted and negotiated on a project-specific case. The drawback of the majority of facilities management contracts is that they do not lend themselves to the flexibility required for the expansion of services.

The shareholder and insurance agreements

These agreements are between the SPV and the various finance providers. They contain terms such as obligations to repay capital and interest, the maintenance of defined debt cover ratios and other covenants and the provision of information on financial and operational performance. Insurance is from either bank loans or bonds from institutions, while cover for insurable risks is borne by the insurance market (PricewaterhouseCoopers 2003). Fight (2005) states that shareholder agreements' contractual framework will need to cover the following (Fight 2005):

- mechanistic provisions;
- interest rates and provisions;
- lender protection against increased costs and illegality;
- representations and warranties;
- events of default;
- miscellaneous provision.

Readers can consult Yescombe (2002) and Fight (2005) for further details on the above issues.

Direct agreement

This contract regulates the relationship between the public sector and the lenders, as the loan agreement is financed by the cash flows arising from the supply of the service (PricewaterhouseCoopers 2003). This is indispensable and deliberately designed to enable the lenders to the project to negotiate with the public sector procurers for the replacement for an operator if the operator is removed, in order to prevent the forced liquidation of the SPV and the loans. All PFI/PPP projects' key contract counterparties are required to sign this agreement. The issues that may be included in such an agreement are described by Yescombe (2002).

Contract governance

Traditional contract methods have provided construction stakeholders with mechanisms of governance based on the functional hierarchy of the existing procurement methods. In a traditional governance structure, the clients govern the service/asset development and its cash flow. The PFI/PPP-based procurement governance requires a new approach to the planning, management and control, which addresses the unique and sometimes complex relationships between projects' stakeholders. Appropriate and adequate governance arrangements are essential for the success of PFI/PPP projects. The aim here to set clear mechanisms to support the operational control processes and manage the interface between PFI/PPP projects' counterparties. The role of project governance, as outlined by Turner and Keegan (1999), is to do the following:

- set strategic objectives;
- set and monitor levels of performance, especially profitability and service quality;
- provide finance, and control financial returns;
- provide technical expertise;
- provide an audit function;
- control risks.

A contract's governance schedule should set out procedures for dealing with these issues. Figure 4.3 illustrates the governance structure for a project-based organisation delivering PFI/PPP projects; further details were discussed on p. 55 and more details are given in Chapter 5.

It is important to point out that there is not one structure for the organisation of PFI/PPP projects, but different governance structures for every project. In most cases, a hybrid structure is adopted. The SPV board maintains relationships with the client, lenders, consultants, and subcontractors directly. Clients set the strategic objectives of the output or service specifications. Lenders provide finance, and create project-specific governance systems to protect their interests. It is stated that debt and equity ownership provides critical monitoring of managerial actions of an SPV and its subcontractors (see Chapter 5). The ownership of this will also give the lenders the ability, in some cases, to hire and fire directors, and approve important operational decisions (Esty 2003b). Technical expertise is normally provided by a pool of consultants and contractors in a subcontractual relationship to the SPV. The audit function is carried out by both the SPV and the clients in conjunction with appointed independent consultants. Control of risk is usually addressed through contractual mechanisms; each party controls their own risks. The operational contracts are often governed through positive and negative incentives. Penalties are incorporated into the contract if the contractor fails to provide the required services (see Chapters 10 and 11).

Summary

This chapter has reviewed the characteristics of the PFI contract system. The PFI contract consists of several interrelated 'agreements'. These are construction, facilities management, and direct and insurance agreements. The content and terms of each contract vary from project to project. However, in the UK, these contracts are standardised. PFI/PPP contract governance is exercised through the creation of an SPV company, risk transfer and contract clauses. Incentives are also used in the operational management of PFI/PPP deals.

Chapter 5

Financing PFI/PPP projects

Introduction

This chapter seeks to explain the PFI/PPP project finance environment, the mechanisms for project funding and the modalities of payment. The majority of PFI/PPP projects are financed through project finance mechanisms. The market for project financing has progressively matured in the past decades and is influenced by various factors. Project finance is a funding mechanism that takes into consideration the credit status of a company, but on cash flows that will be generated from sponsored projects. Risk transfer lies at the core of project finance which will be discussed in Chapters 12 and 13.

PFI/PPP capital project environment

PFI/PPP initiatives have grown steadily as fiscal resources have become more constrained. The rationale behind the approach of PFI/PPP was already discussed in Chapter 1. The project finance method allows a suitable division of roles among the stakeholders involved and it is designed to ensure that projects will be designed, constructed and operated with a minimum risk over the life cycle of the asset. In project finance, the environment generally depended on the complexity and size of the projects. Projects involving less than a billion sterling are well within the capacity constraints of most lenders. Large projects may have to be divided into smaller projects in order to secure private funding or alternatively to secure government funding or guarantors.

PFI/PPP projects are usually financed and delivered by a consortium. The consortium's main purpose is to finance and deliver the service/project to the public sector based on value for money principles (see Chapters 1, 2 and 3). How the consortium is organised depends largely on the size of the project and the modality of financing adopted. The PFI/PPP project environment depends on the contractual structure of the project in question, risk allocation, capital structure, and the reliability of forecasted revenues. A typical PFI capital environment is shown in Figure 5.1, which displays the relationship between different entities in a typical PFI/PPP project market. This constitutes the first

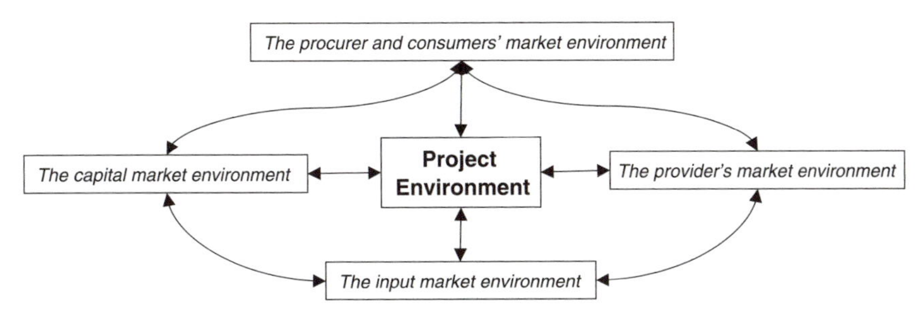

Figure 5.1 PFI/PPP capital project environment.

step in understanding the risk-bearing arrangements crafted between different parties in the procurement of PFI projects and the capital financing required to sustain those risk structures. Figure 5.1 illustrates how a typical project environment consists of four interrelated project market environments:

- *The capital market environment* in which PFI/PPP projects' investments are financed. The participants in a typical PFI financing structure include equity providers, debt providers, loans, bonds (some of the loans and bonds are index linked) and mono-line insurance providers.
- *The provider market environment* in which intended services are delivered based on the value for money paradigm. It is the provider's responsibility to design and construct building assets in response to the requirements (output specification) set by the procurer. The provider also finances the initial investment and delivers a range of specified services throughout the concession period.
- *The input market environment* includes contractors, suppliers and all tangible and intangible resources that are necessary to deliver the required services. The relationship between the SPV and the contractors is governed by additional contracts for a specific period of the project life cycle. The relationship between contractors, subcontractors and suppliers will also be governed by additional financial contracts.
- *The procurer and consumers market environment* specifies ongoing requirements for services and facilities. Also, it pays a single (unitary charge) payment to the provider to cover both the cost of financing and the services. After the concession period the asset normally reverts to the public sector. The end user or consumer drives the demand at all the various stages in the value chain of the PFI/PPP project market environment.

A PFI/PPP capital project environment encompasses the whole contractual structure of project financing, including, for example, ownership structure or operational and financial risk management and is thus not simply a funding contract. As such, the venture's business risk is shared by sponsoring owners, contractors, host governments, suppliers, customers, and even creditors, who commonly would not bear equity risk.

Organisational structure of project finance

Investments in projects require financing and this can come from a variety of sources. The project funding is organised in several ways. PFI/PPP projects' funding structure is usually dictated by tax and existing legal framework, as well as by the creditworthiness and implications for each participating organisation. It is important that the legal framework and structure of the project funding should be established on the basis of maximum flexibility to allow the establishment of other contractual independent legal entities such as construction, operation and refinancing of projects.

Project finance companies have evolved as institutional structures that deliver value for clients through better financial management of resources, risk management and better utilisation of resources through the life cycle of investments. One of the most important structural attributes of project funding structures is the distinction between the project (asset) and the financing structure. Project funding can be obtained from various sources. As a foundation for the theoretical understanding of the organisational structure of different project funding mechanisms, this chapter begins by describing and presenting the difference between public, corporate and project funding structures. The comparison is based on five characteristics developed by Esty (2003b). These are: organisational structure, capital structure, ownership structure, board structure and contractual structure.

Public finance structure

Figure 5.2 illustrates the structure of funding projects using public finance. In this mechanism, projects are funded from the existing balance sheet, surplus funds or issued debt (government bonds) to be repaid over a specific period (Manual 2001). The attributes of this method are presented next.

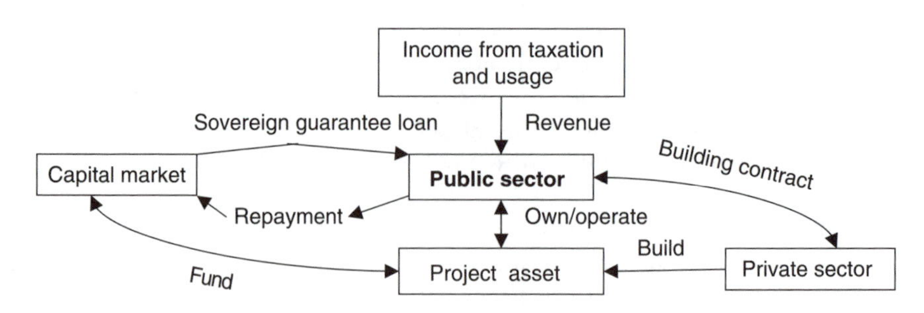

Figure 5.2 Public finance structure.

- *Organisational structure*
  The finance and project structure are indistinguishable. The public sector is responsible for financing, developing and operating the asset. There is no separate contractual arrangement for financing projects.
- *Capital structure*
  Public institutions borrow funds to finance projects. Usually the government gives a sovereign guarantee to lenders to repay all funds. Project funding might be raised from a combination of equity and debt. The debt will be shown as a liability on the government's balance sheet.
- *Ownership structure*
  A project's public funding has highly concentrated borrowed funds, usually from government bonds, and its own equity ownership structures. The public sector (the government) is liable to pay all the creditors. Usually before leading funds creditors analyse the government's total ability to raise funds through the normal taxation mechanisms. The level of interest and risk depends on the level of equity ownership by public institutions.
- *Board structure*
  Project boards are comprised primarily of public institutions.
- *Contractual structure*
  Here public institutions are responsible for all contracting and subcontracting throughout the life cycle of the asset, including construction and operation stages.

Corporate finance structures

This type of project finance structure is associated with private sector institutions using their own capital for funding projects. This mainly applies to projects that are limited in size and do not require outside financing. This is due to the fact that private companies are able to absorb this financial commitment in their balance sheet. However, most corporate institutions tend to avoid this method because it has the potential to limit their participation in future projects (Manual 2001). Figure 5.3 illustrates the organisational structure of the corporate finance structure and its organisational attributes are explained in the following subsections.

- *Organisational structure*
  Project finance is not a separate incorporation of the private company. Under this scheme the private company's existing structure would govern the project/asset and its cash flows. By doing so, private companies could hold large risky assets on their balance sheets.
- *Capital structure*
  A private company may choose to use its own equity or to borrow funds to develop the asset in question and guaranties to repay lenders from its available operating income and its capital of assets (Manual 2001). Loans

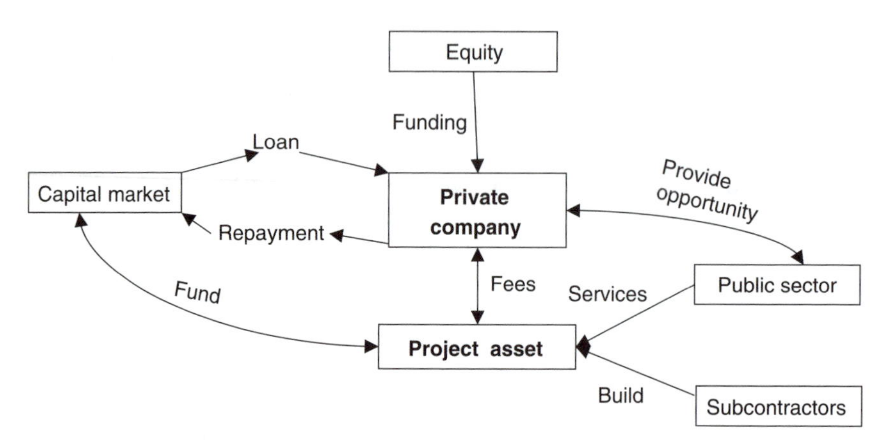

Figure 5.3 Corporate finance structure.

under this system are based on the recourse debt method. Each project has its own characteristics that may alter the standard format of the optimal capital structure. Guzhva and Pagiavlas (2003) suggested that corporate capital structure is influenced by variables such as fixed assets, no-debt tax shield, investment opportunities, firm size, volatility, probability of bankruptcy and the uniqueness of the product or service offered by the firm.

- *Ownership structure*
  Most the funds are generated from the private company's assets and loans from lending institutions. The company is liable for debt repayment to creditors. All loans will be shown on the balance sheet as a liability to the company.
- *Board structure*
  Under this plan the project boards are comprised primarily of the private company board of directors and possibly shareholders, depending on the size and complexity of the project and its finance.
- *Contractual structure*
  The private funding company is responsible for construction, operation and subcontracting all related risks.

Project finance structure

Project finance is defined as the creation of a legally independent project company financed with non-recourse debt for the purpose of investing in a capital asset, usually with a single purpose and limited life (Esty 2003b). Project financing uses the project's future revenues as the basis for raising capital to fund the construction and operation of an asset over a specified period. The system is based on the paradigm that the private sector uses its resources to build, maintain, manage and operate public assets. It is also stated that the use of management skills and know-how of private enterprises will lead to better public services at a lower cost, a view that has been contested by others who oppose the idea of private funding. Project financings are extremely complex. It may take a much longer time to structure, negotiate and document a project financing than a traditional financing, and the legal fees and related costs associated with project financing can be very high.

The legal organisational structure of each independent project company varies, dependent on the size and type of projects. Figure 5.4 illustrates the unique organisational structural attributes that characterise a typical legally independent project company. These unique features are described in detail in the following sections.

- *Organisational structure*
  As stated above, projects delivered under the private project finance system are rarely financed and delivered in their entirety by a single provider. It is common that a consortium is formed to deliver the assets and associated services. The input from the members of the consortium is managed through a special purpose company which is a separate legal incorporation from the members' organisation. SPV is a legal entity in its own right. The creation of the SPV organisational form amounts to risk management via organisational form. The structure of SPV is dictated by tax and legal considerations, as well as by the credit implications of each member (Manual 2001). The reasons SPV has more than one sponsor are attributed to the following (some are adopted from ibid.):
  - financial and technical requirement exceeds capacity of the sponsor;
  - risk sharing;
  - sponsors complement each other's capabilities;

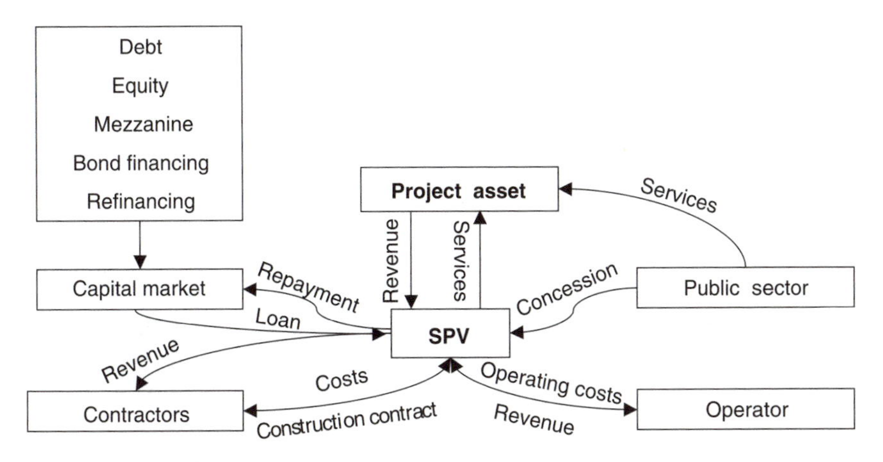

Figure 5.4 Project finance structure.

  - bankruptcy remoteness, that is, to prevent the project from a potential bankruptcy of any of the providers.

- *Capital structure*
  The SPV is capitalised up to 90 per cent with senior debt provided by a sponsor or bank to an agreed schedule of capital and interest repayment. The remaining 10 per cent is capitalised with equity contribution from each of the sponsors or from some of them in the majority of cases. This funding is non-recourse beyond the SPV. This is very important in the unlikely event that the SPV goes bankrupt; the sponsor has no rights to recover any losses from any other party. The SPV uses very high gearing (the level of debt to equity) compared to a corporate finance structure.
- *Ownership structure*
  An SPV has highly concentrated debt and equity ownership structures. Most of the debt is from financial institutions and is non-recourse to the sponsoring firm. Lending to the project is non-recourse to the providers/sponsors by virtue of the limited liability nature of the SPV. As a result, creditors look to the project's future revenue stream generated by the project to repay all debt. The public sector does not provide a financial guarantee to lenders.
- *Board structure*
  SPV boards are comprised principally of affiliated directors from the sponsoring organisations. The complexity and size of the boards depend on project size and the complexity of the financing deal.
- *Contractual structure*
  The SPV is created with the aim of undertaking a defined activity. Execution of the activity may require the involvement of a number of parties. The SPV will enter into subcontracts with other organisations for the execution of these activities. There are four major project contracts that govern the supply of necessary resources, construction, operation and maintenance of the asset throughout the concession period. Complex projects can have tens of subcontracts. Arrangements for financing the SPV are normally concluded at the same time that the contract and subcontracts are entered into.

Credit guarantee finance

Credit guarantee finance (CGF) is a new financing mode proposed by the UK HM Treasury to meet the investment challenge. Under this scheme, projects are funded by central government (raising cash through government bonds) with a repayment guarantee from a private sector or mono-line insurer/assurance. The system works on the basis that the public sector will provide funds to the projects by way of cash advances under the terms of a loan agreement between the public sector and the private sector. Loans are repayable after completion of the project. Loan repayments are made according to a strict financial model and repayment schedule agreed with stakeholders, i.e., the SPV, procuring authority and guarantor. In exchange for providing this loan capital, the government will receive an unconditional repayment guarantee from the guarantor (HM Treasury 2003a). The theory behind this framework is that the Treasury should fund the PFI/PPP projects while banks and mono-line insurances would simply guarantee that the Treasury would get its capital back if the project fails to perform to its original expectation. Other perceived aims of the scheme are to reduce the margin paid on loans in PFI/PPP projects, which is alleged as being too high by the government. Another stated benefit is that the government would not match funding for its investments, hence there would be no negative carry-costs associated with advance funding during construction. The aim of using such a system is to ensure a balance of guarantors between the mono-line insurers' banking and other stakeholders, with clear restrictions on exposure of the public lenders as a whole to each subgroup. It is suggested that under this method there are several factors that expose banks in particular to extra risks such as credit deterioration issues relating to syndication of risk and liquidity considerations. If such a risk is reflected in pricing, it would probably wipe out the initial gain from loans payment. It is also possible that if sponsors have no input to the selection of the guarantors, it can lead to a direct impact on bid costs which could lead to a high yield requirement for equity and subdebt compared to the normal PFI/PPP financing models described above.

Sources of project finance

It is well known that one of the main obstacles in the delivery of public and private projects is the availability of funding. Potential sources of project financing include commercial banks, leasing companies, insurance companies, pension funds, governmental bond authorities, finance companies, export credits, international financing agencies, private lenders and customers (Farrell 2003). A project's participants can also be a potential source of project finance. This section outlines the types of long-term sources of capital that a project may use in raising funds. Sources of capital for the development and operation of PFI/PPP projects include a combination of senior debt, mezzanine debt, equity and public reserves. Each source of financing will take differing levels of risk, and have different financing terms attached to them.

Debt

Debt represents the predominant vehicle through which firms raise external funds. The market for project debt is influenced by various factors, including the contractual structure of the project, risk allocation and the reliability of revenues (Treharne 2003). Debt capital falls into two categories, as shown below.

1 *On balance sheet finance*: This lending is provided by commercial banks or the debt capital markets against the level of sponsors' corporate covenant. This type of funding is only appropriate for financially strong project sponsors, who are able to absorb the associated liability and risk. Having said this, such a financing system is simpler and less costly to deploy than senior debt. However, repayment of the debt is not based on any asset or performance of provided services but is secured against the assets and future earning capability of the sponsor's whole business. It will appear on the balance sheet of the sponsor as liability.
2 *Limited recourse finance*: At its limit, this lending is referred to as senior (non-recourse) debt where the bank funding is non-recourse beyond the SPV, that is, in the event of the SPV failing, the bank has no rights to recover losses from any other party. However, in practice, lenders may require sponsors to cover some of the risks by providing guaranties. Lenders will also place a number of controls including the implementation of cover ratios and a limitation on the disbursement of cash from the SPV. This lending is provided by commercial banks to an agreed schedule of capital and interest repayment. This type of debt is senior to equity in receiving dividends and repayment of principal. Raising senior debt is relatively cheaper and accounts for 90 per cent of the capital cost of PFI/PPP projects in the UK.

Equity capital

Equity funding is more expensive (in terms of cost of capital) than debt, but can often be simpler and less costly to deploy. This is the most risky form of finance. Equity funding is provided by investors or in some cases the parent company of the construction and FM provider. Equity investors bear a greater level of risk but require a higher return than that required by senior debt providers. Equity debt returns are repaid after senior and mezzanine debt payments, and tend to arise towards the end of the concession period.

Mezzanine debt

Third-party equity is available from both investors and commercial banks usually in the form of subordinate or mezzanine debt. This is a more expensive form of finance. Mezzanine debt has fewer rights than senior debt. In the event of SPV failure, it would have a lower level of rights over the asset. However, it can be used to give lenders a stake in the SPV and align their interest with the equity sponsors.

Bond financing

A bond is a loan, shares in which can be traded on the stock exchange, just like shares in equity. There is increasing use of the bond markets to fund PFI/PPP projects. In PFI/PPP projects, interest-paying bonds are issued by the SPV which are then sold on the financial market. This type of funding is mostly used to finance large projects. Usually bond financing is used as an alternative to senior debt. In this funding mechanism the payments from the return from the project are used to pay the interest on the bonds. It is a requirement that the bond issued must be underwritten by a lender so that the SPV has a guarantee that it will have adequate cash to carry out its activities. Generally, these types of bond are acquired by financial institutions such as pension funds and insurance companies, who have long-term financial obligations and therefore appreciate owning long-term financial assets. Accordingly, a particular attraction of a bond relative to a bank loan for a long-term infrastructure project is the much longer maturity that can be achieved.

Refinancing

Project refinancing is developing very fast and secondary trading is also current. The key issue that influences the refinancing process is how to divide the financial benefit of refinancing between the private and the public sector. In the UK, this problem has been resolved, with a mandatory 50/50 profit split on all new transactions and a 70/30 voluntary split on existing ones. This sharing can occur via payment to the public sector or through reduced service charges. The secondary market is finance-driven, not construction-driven and could lead to offering investors high returns and boosting liquidity. This is an excellent opportunity for PFI/PPP investors to maximise return on their investment by refinancing debt, after the risky construction stage of a project, securitising project revenues or even equity dividend streams, or ultimately selling to a secondary investor who wants to hold long-dated assets. According to Fitch Ratings (2003), refinancing may entail higher gearing than the original transaction, as a result of the maturity of PFI/PPP market, the absence of construction risk, and the availability of longer-term finance at lower interest rates.

Financing strategies

Financing strategies depend on current banking market conditions, the level of competition in the financial market, and the characteristics of the project and its overall risk profile. Terms and conditions of the project debt initially will depend to a large extent on maturity, the purpose of the project, or the likelihood of default by any stakeholder. Generally, however, investors aim at maximising their returns while minimising their risk. A project consortium aims at minimising the level of equity in favour of debt. Contractors and operators can own a percentage of the project company, sometimes up to 100 per cent. Although they are equity investors in the project, their primary interest is in the earnings they will generate in construction operating and maintaining of the project.

The rule of thumb is this: the higher the level of debt to equity, the more financially risky the project will be. The lower the level of debt to equity, the less profitable the project will be. The margin of profitability is very difficult to determine accurately because the profits are shared by a large number of equity investors in the project. Hence, finding the right balance between equity and debt is always a difficult issue. A project consortium must balance the higher returns on equity against greater exposure. However, since debt usually comes at a lower price, the consortium must balance the amount of debt against the overall financial costs of the project. The conventional practice in UK PFI/PPP projects leans towards a debt equity ratio of 90:10. Issues that are addressed in developing project financing strategies are discussed in the subsequent sections.

Pricing project finance

Pricing project finance can be volatile and unpredictable and it is dependent mainly on the risk package on a particular project and to some extent is related to market fluctuations. Pricing here refers to the cost of borrowing the funds. Pricing project funds is usually based on existing market rates plus an amount for risk. The real cost of borrowing is the inherent risk of the project. The cost of financing to the private sector finance has the element of risk built into it, whereas public funding of a project is underpinned by government funds. As described in the previous section, some of the funding sources require higher interest rates than others. PFI/PPP projects debt pricing depends on the following factors:

- legal framework and guarantor support
- construction risk
- credit risk of the payer
- market risk
- added value by sponsor
- project performance risk
- contractual risks.

Lenders will consider all the above issues in evaluating investment decisions, the level of risks to which they will be exposed and, ultimately, the rate of return required in pricing project finance debt. It is also safe to state that commercial lenders will address the above risks by simply increasing loan pricing or raising the level of coverage ratios.

Financing parameters

Financing parameters or covenants are used to control the borrower and to ensure that its financial condition has not materially changed from that which the lender anticipated when making its initial credit assessment and deciding to grant the funds. Financial parameters are initially derived from the projected cash flow but once a project is operational they are derived from the actual project, balance sheet and the income and cash flow statements. The cash flow projections must be sufficient to service any debt contemplated, provide for cash needs, pay operating expenses, and still provide an adequate cushion for contingencies. Lenders often look at these projections to determine debt coverage over the loan life. Project finance feasibility studies examine various critical financing parameters; the most common are annual debt service cover ratio (DSCR), loan life cover ratio (LLCR) and project life cover ratio (PLCR). The equity investors will also examine the project's projected economic performance. Their key measurement is usually the internal rate of return. Such parameters are included in most project finance agreements and are outlined below:

- *Leverage ratios*: financial leverage ratios provide an indication of the long-term solvency of the project and measure the extent to which the project or SVP is using debt.
- *Debt service coverage ratio (DSCR)*: a measure of the ability of a project to service its debt, this measures the relationship of annual cash flow to the amount of debt outstanding. DSCR is one of the key indicators of the creditworthiness of the project. DSCR is the ratio of cash flow available in any period to the level of cash needed to cover the debt repayment (principal and interest). DSCR is defined as the earnings before interest, taxes, depreciation and amortisation divided by the debt service (the payment of interest and repayment of principal). In a backward-looking covenant, DSCR measures the cash flow available to meet debt service for the previous period, to the amount of actual debt service for the relevant period. In a forward-looking covenant, DSCR will determine how much principal the borrower is obliged to pay on any given repayment date. DSCR is measured on an annual and loan life basis. The purpose of such a ratio is to require the borrower to repay such an amount as to ensure that after repayment, the given DSCR would continue to be met. DSCRs vary through the life of a project, depending on the capital, operating and maintenance cost programme and on the ability of the project to generate revenue. DSCRs are usually lower in a project's early years. However, the acceptable threshold value of DSCR varies with the risk profile of the project and ranges from 1.35 for a low risk project with a high degree of contractual certainty to between 1.7 and 2.0 for projects with volatile revenues or an element of market risk (Maunder and Haggard 2003). Lenders usually take into account the trade-off between cover ratios and gearing levels (the level of debt to equity) where the level of DSRC reflects lenders' risk perception of the cash flows themselves as influenced by aspects of the financial contract such as contracted revenues, revenue variability, asset operation, maintenance costs and all other unidentified

risks. The idea here is that the level of debt to equity (gearing level) is a function of the minimum DSCRs required by lenders. In other words, the lenders aim at minimising the level of equity in favour of debt for reasons explained in the previous section. DSCR is defined as follows:

$$\text{DSCR} = \frac{\text{EBITDA}}{\text{interest and principal charges}}$$

Where EBITDA = Earnings before lease charges, interest, taxes, depreciation and amortisation.

- *The debt to equity ratio*: the debt to equity ratio indicates how much the project or SPV is in debt or leveraged and provides a window into how strong the project finances are. Lenders and investors use the relationship between equity and debt to evaluate financial risk. A high debt to equity ratio indicates that the project may be over-leveraged, and also the project is financially risky. If the level of debt to equity ratio is low, this implies that the project might generate less profitability to investors due to the fact that the profits are shared by equity investment in the project. The ratio of debt to equity varies according to the characteristics of the project and the volatility of cash flows. The debt to equity ratio is computed as:

$$\text{debt to equity ratio} = \frac{\text{total debt (liabilities)}}{\text{total equity}}$$

- *Debt ratio*: debt ratio measures the extent to which the project is using long-term debt. It is acknowledged that when debt shrinks and income increases, the cash flow of the project will improve, but if the opposite is the case, the cash flow of the project will deteriorate. Projects with large long-term debt carry a huge burden of interest payments, and a risk of not having enough working capital to operate the project. This is why the debt ratio is used by investors to set up the optimum leverage of a project. According to Esty (2002), the average debt ratio varies between 70 and 90 per cent, depending on the characteristics of the project to be financed. It is also true that debt investors aspire for a lower debt ratio for the reason stated previously. The debt ratio is defined as follows:

$$\text{debt ratio} = \frac{\text{total debt}}{\text{total asset}}$$

- *Loan life cover ratio (LLCR)*: the loan life cover ratio is also one of the key indicators of the creditworthiness of the project. The LLCR takes account of the time value of money, and uses discounted values at an appropriate interest rate. LLCR compares the net present value (NPV) of net cash available over the remaining loan life until the final repayment date with the total loan at the time of testing. This is the ratio between the net present value of future cash flows over the life (maturity) of a loan and the loan outstanding. In reality, LLCR measures how many times over the outstanding loan could be repaid. By computing the NPV of the stream of future cash flow, the level of security available from the income stream cam be established. Most banks have a requirement for an LLCR of around 1.4:1.0 or 1.5:1.0.

- *Project life cover ratio (PLCR)*: PLCR is the value of a project's cash flow available for debt service until the end of the project divided by the principal outstanding. PLCR compares the net present value of cash available over the remaining project life with total loan balances at the time of testing and gives the sponsors reassurance as to their likely value of return. The PLCR value is usually greater than the LLCR.
- *Liquidity ratios*: these ratios are used to assess projects' ability to meet short-term obligations in terms of debt repayment. They show a great insight into the present cash liquidity of the project or SPV and its ability to remain solvent in the event of adversities. Basically, these ratios assist financial analysts to compare a project's short-term obligations with the short-term assets available to meet these obligations. Liquidity ratios for the purpose of this chapter are grouped into two types: current ratio and acid-test ratio (sometimes referred to as quick ratio). Current ratio is the most frequently used of these ratios. The current ratio measures the ability of a project's current assets in terms of cash, inventory, and receivables to cover current liabilities that is short-term (usually one year) debt, and accounts payable. The current ratio is the ratio of current assets to current liabilities:

$$\text{current ratio} = \frac{\text{current assets}}{\text{current liabilities}}$$

The higher the ratio is, the greater the ability of the project to pay its liabilities. The current ratio does not capture the liquidity of all components of the current asset. A more accurate guide to liquidity is the quick ratio. The acid ratio is the same as the current ratio, except that it excludes liquid assets or inventories. This might be due to that fact that inventories are the least liquid component of the current assets.
The quick ratio is defined as follows:

$$\text{quick ratio} = \frac{\text{current assets} - \text{stocks}}{\text{current liabilities}}$$

The ratio measures the availability of cash, marketable securities, and receivables in relation to current liabilities and that is why presumably it provides a more informative measure of liquidity than the current ratio.
- *Profitability ratios*: profitability ratios are used to capture or measure the project's operation efficiency and offer different measures of the success of

the project at generating profit. They are of two types: those showing project sales, and those showing profitability in relation to investment. These ratios are mainly of interest to the project sponsors and equity investors due to their usefulness in showing the level of profitability that can be expected from a given investment. Normally these ratios are used by regulators as the basis for determining changes in a regulated project or payment by the government in a publicly funded project (Manual 2001). The return on assets ratio is a measure of how effectively the project's assets are being used to generate profits, whereas the return on equity ratio measures the profits earned for each amount of cash invested in the project by equity holders. The ratios are computed as follows:

$$\text{return on assets} = \frac{\text{net income}}{\text{total assets}}$$

$$\text{return on equity} = \frac{\text{net income}}{\text{investor equity}}$$

- *Internal rate of return (IRR)*: the IRR measures the return to equity on the money invested. The return is the compound return earned based on the initial investment and the future cash inflows. IRR is the discount rate which equates the NPV of the cash flows to zero; methods computing IRR and NPV can found in the work by Boussabaine and Kirkham (2004) or any other appropriate resources. The return on equity and IRR ratios are similar and their values can be from 10 up to 30 per cent after inflation dependent on country, and the nature and risk of the project.

The above financial parameters or ratios will assist in making rational decisions in evaluating the performance of PFI/PPP projects. Analysis and interpretation of the above ratios should give project stakeholders a better understanding of the financial performance of the project. However, ratios by themselves can be misleading. That is why they should be used as indicators, and they must be used in combination with other measures before reaching rational or optimum decisions. It is also important to stress that ratios are highly dependent on the limitation and accounting practices. Different practices may result in completely different sets of ratio value.

Under the PFI/PPP schemes, providers are responsible for developing the financing strategy of the project. Usually financing strategies are based on the cash flow requirements of the project using some of the sources of funding and financial parameters described above. In addition to these techniques the following financial strategies might be used to optimise cash flow and return on investment:

- Extending the debt term leads to lower payments and increases DSCR, plus it does present more risks to lenders and reduces cash flow requirements.

- In some circumstances, especially in developing countries, one option is to defer principal payments, which will increase the DSCR ratio but the lender risk will increase accordingly. These strategies are necessary because the cash flow returns from the project are insufficient to meet debt service requirements.
- Meeting project cash flow requirements through borrowing with an agreed grace period. This will have the consequence of increasing the debt service and the DSCR ratio and undoubtedly it will lead to a higher interest rate.
- Reducing the debt service through third party creditworthiness. This will have the effect of reducing interest rates but it may lead to an increase in the third party contingent liabilities.
- Increasing equity share in the project finance will reduce total debt as well as reducing principal and interest payments and return on equity. However, it will increase DSCR ratio, taxes and profit.
- Shifting some the financial risks to subordinate lenders through mezzanine methods, although an expensive way of financing projects, will improve DSCR for senior debt institutions and may create the right conditions for investment in the project. However, this financing strategy has the consequence of increasing total debt service costs and results in higher interest rates for subordinated debt.
- Maximising the use of senior and mezzanine debt through lower development cost and lean debt structures.

Payment mechanisms in PFI/PPP projects

The payment mechanism is one of the most important agreements in the PFI/PPP procurement process. In PFI/PPP contracts, the provider does not receive a payment until the facilities are available for use. The quintessence of PFI/PPP procurement is the provision of a specified quality service. Hence, the payment is linked to performance, availability of assets, quality of services provided and sometimes to level of use. Thus, payment arrangements are seen by the public sector as an effective risk allocation tool.

Revenues received from the provision of service are apportioned between capital repayments and operating revenue. Payments for services may be adjusted based on prevailing market rates every five years. Once the construction phases are complete, the service payment usually varies only for the impact of the indexation mechanism. The retail price index (RPI) with a benchmark base rate can be used for this purpose. It is also possible that the unitary payment will be subject to inflation on an annual basis to the end of the concession. The inflation of the unitary payment can be based on the RPI or fixed inflator provided in the contract, or a combination of both. The payment arrangements are determined not on the basis of payments for particular inputs, but instead on the basis of one or more of the following criteria:

- availability of service;
- performance standards;
- usage of the service by a third party;
- payments might be linked to improvements made in the relevant service.

Service-based payments are particularly suited to projects requiring accommodation services, such as prisons, hospitals and schools, from which the procurer may deliver core services. Here, availability is the most important factor for the continual running of the service. The usage may be variable but it is not as important as availability. Usage as a basis for payments is hardly used on its own due to its crude nature of measuring public services. However, it can be used as yardstick for measuring the attractiveness of the service and may implicitly reflect service quality. Benefits-based payments might be relevant in some circumstances where the public sector requires improvements in services such as education standards or environmental performance.

The unit service delivery charge will be fixed for the term of the project agreement, subject to adjustment for inflation and changes to the services provided that the procurer may require. The unitary charge is also subject to market testing at predefined intervals (see Chapters 10 and 11).

An effective payment system should be structured in a way that will encourage the private sector to deliver efficiency savings over the whole life cycle of the project, with such savings to be shared between the procurer and the private sector. One key feature of payments under PFI/PPP schemes is the consideration of reducing or withholding payment of service delivery fees for defaults in providing the specified services to the standard in the project agreement (for further details, see Chapters 10 and 11).

The UK Treasury's (1999) technical note on *How to Account for PFI Transactions* identifies three broad models for payment mechanisms:

1 *Model A*: this model is based on payment for the provision of a number of available places. No payment is made for unavailable places. For a place to be declared available, both the physical space and the associated core services must be available. Deductions from the unitary payment can be made for substandard performance of contracted services. This payment structure is seen as non-separable (i.e., there are no separate payment streams for any of the non-core services not contracted or associated with the definition of available space). The payment scheme is represented by the following formula:

   $P = (f + i) - z$
   Where:
   P = unitary payment per place
   F = fixed amount per available place per day

I = indexed amount per available place per day (based amount increased by RPI)
Z = performance deduction

2 *Model B*: this model is based on a unitary payment on the full provision of an overall accommodation requirement which is divided into different elements or units. There are no separate payment streams for any of the non-core services. If the service provider fails to provide an available element, there is a payment deduction equal to a proportion of the unitary charge which depends on the importance of each element. Availability is usually related to a unit being usable and accessible according to certain criteria defined in the contract relating to minimum standards of service provision. Availability will cover core services, and sometimes non-core services, such as cleaning and heating, within a specified standard. Inferior performance leads initially to performance points and to payment deductions once a certain level of points has been accumulated. The ceiling of these points is usually negotiated during drafting of the payment contract (see Chapters 10 and 11). This payment structure is seen as non-separable and is represented by the following formula:

$P = (F^*I) - (D + E)$
Where:
P = unitary payment per day
F = price per day for overall accommodation requirement
I = indexation factor
D = deductions for unavailability
E = performance deductions.

3 *Model C*: unlike models A and B above, this payment scheme is a combination of an availability payment stream and a performance-related facilities management payment stream. This type of payment structure is separable, i.e., the individual elements of the PFI payments operate independently from each other, with deductions against the performance payment stream for failure to meet the required performance levels. Availability is defined on a similar approach as in model A and B, and areas are divided into units and attributed a different deduction percentage for unavailability according to their importance. The following formula is used to represent the structure of the payment:

$P = A + Q - D - E$

Where:

P = unitary payment per unit
A = availability payment
Q = indexed facilities management payments
D = deductions for unavailability
E = performance deductions.

It is clear from the above models that PFI/PPP performance is directly linked to the payment schemes. The aim here is that the client relies on performance as the basis for payment to ensure that appropriate risks are transferred to the private sector. The performance systems envisaged under PFI/PPP contracts are based on the output specification. Services output specification and a performance measurement management system are used to prescribe the level of performance required. These procedures provide the level of service required by the client in each aspect of the contract scope, the priorities for service availability and responsiveness and standard of service to be achieved (McDowall 1999). Throughout the period of the concession the payment is adjusted to reflect the performance of the provider. Figure 5.5 illustrates how a typical PFI payment structure is organised around the output specification. The figure illustrates that if performance is sub-standard, then service failures arise, and this will result in deductions from the service payment. The service payment is based on asset, function and service output specifications. The asset output specifications refer to the requirements for design and performance of the

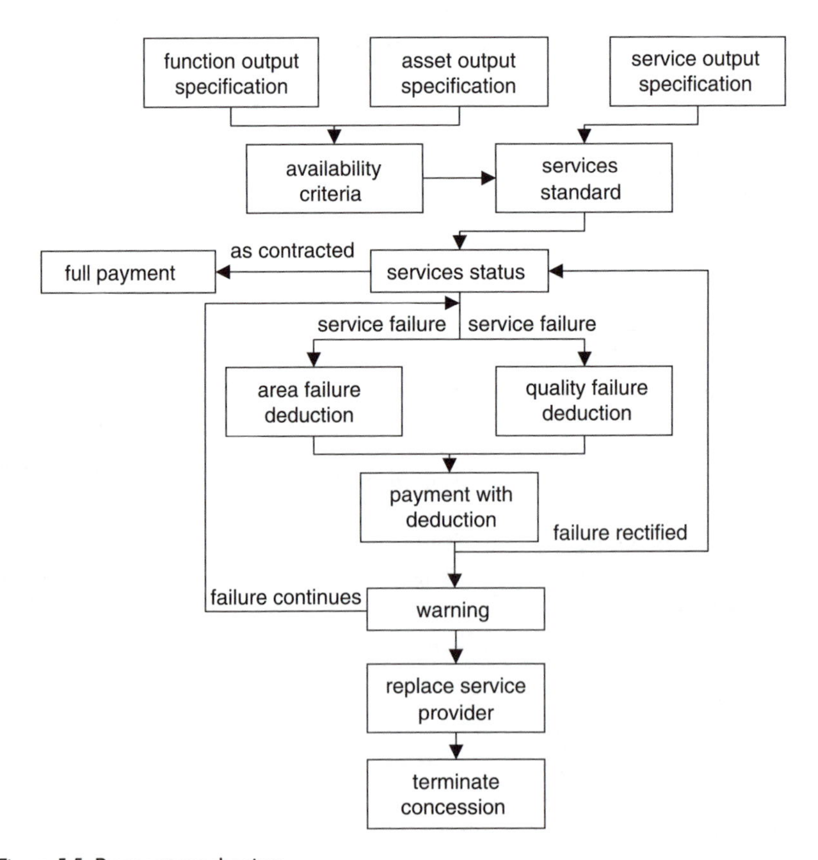

Figure 5.5 Payment mechanism.

building fabric and services and provide use parameters for each functional area setting out the performance standard in aspects such as temperature, humidity, air flow, lighting, safe water, safe sewerage, structural performance, etc. The function output refers to the requirements for each of the functional areas, such as classrooms, staff rooms and music classrooms. In health projects, this may refer to the clinical purposes for the area. The criteria specified in the asset and functional output specifications form part of the availability criteria for each functional area. Usually these two aspects are pulled together in room data sheets as shown in Table 5.1. It is through reference to data in Table 5.1 that the specific availability criteria for any area can be determined. Under PFI contracts, an area is regarded as available if it satisfies each of the following conditions:

- the accessibility condition;
- the safety condition;
- the use condition;
- the prescribed health function condition.

The service output specifications are project-specific and provide details on the service outputs that the provider is expected to deliver. The service output specification might be classified into:

1 *Building design output specification*: This deals with the design requirements for buildings, grounds and accommodation. This is related to the asset management policies and the operational and capacity requirements.

Table 5.1 An example of service performance specification

FM element	*Performance requirements*	*Standard performance*	*Weighting /priority level*	*Rectification time*
Security	Patrolling routes/ work schedules regularly reviewed/updated	Work schedules available for all areas	10/4	5 minutes
Cleaning	Replace consumables in all other areas/clean sanitary areas	Attend and perform the task within required target Response Time	20/2	5 minutes
Maintenance	Response to emergency repair and Response to urgent repair	Make good any damage caused; at the designated quality standard in room data sheets	30/1	Temporary 1 day and permanently in 5 days

2 *Building service output specification*: This comprises all the performance requirements and quality standards for the estates and management service; catering services; caretaking, security and safety services, and the equipment provision and maintenance services.

The service provision has to comply with agreed service standards. Sometimes the whole building is subdivided into priority zones, which are weighted differently for the purpose of calculating the penalties for any availability default.

Accounting treatment of PFI/PPP projects

The key question in accounting for PFI/PPP projects revolves around two main issues. These are:

- Whether a party to a PFI/PPP contract has an asset of the property, i.e. they control the property and should therefore record the property on their balance sheet, and not in the books of the legal owner of the asset.
- Whether a contract should be separated into different elements or regarded as a whole. This issue arises from the difficulty of separating the unitary payments into service and property elements.

The first issue addresses the question of who owns an asset of the property. In most cases, the property is owned or leased by the operator over the full period of concession. At the end of this period, the property will either revert to the procurer at a fixed price, the contract will be re-tendered, or the asset will be retained by the operator. In order to determine who has control over an asset of the property in a PFI/PPP contract, it is essential to identify which party has access to the benefits arising from the property and exposure to the risks involved in it. This concept is based on the fundamental principle of PFI/PPP which is that risk is held by or transferred to the party best able to manage it. For this purpose, the UK HM Treasury has adopted the financial reporting standard (FRS) as its measure of the extent to which risks have been appropriately transferred to the private sector partner under a PFI/PPP contract. FRS considers where the risks and rewards of ownership are, and requires that any asset or liability be accounted for on the balance sheet of the party bearing those risks and rewards. That is why the UK Treasury Note *How to Account for PFI Transactions* (1999) advises that contracts should be drawn up very carefully to ensure that the operator has control over the asset and that he is taking the major risks.

Summary

This chapter introduced PFI/PPP capital project environments. A typical PFI/PPP project environment consists of capital, provider, input, procurer and consumers market environments. A foundation for the theoretical understanding of the organisational structure of different project funding mechanisms is explained. Potential sources of project financing are described. Sources of capital for the development and operation of PFI/PPP projects includes a combination of senior debt, mezzanine debt, equity and public reserves. Each source of financing will take differing levels of risk, and have different financing terms attached to them. The chapter points out that project financing strategies depend on current banking market conditions, the level of competition in the financial market, and the characteristics of the project and its overall risk profile. Payment methods in PFI/PPP are presented. An effective payment system should be structured in a way that will encourage the private sector to deliver efficiency savings over the whole life cycle of the project, with such savings to be shared between the procurer and the private sector. Finally, the accounting treatment of PFI/PPP projects is introduced. The rule is that the asset should be accounted on the balance sheet of the party bearing the risk and reward from the ownership of the asset.

Part III

The process of cost planning of PFI/PPP projects

Chapter 6

Fundamentals of PFI/PPP projects cost planning

Introduction

The detailed nature of PFI/PPP cost planning requires the analyst to have at their disposal a variety of mathematical and analytical skills to perform the task effectively. The complexity and size of the project will normally dictate the process and techniques required to cost these projects. This chapter provides an insight into some of the fundamental principles in costing PFI/PPP schemes. The chapter begins by introducing time value of money concepts. Next, the application of discounting techniques and other capital investment principles are introduced. The chapter then moves on to consider the risk in time value of money decisions. Methods for economic evaluation and financial models are also presented. Finally, the chapter explains how project cash flow is developed.

Time value of money concepts

The time value-based investment concepts are employed to evaluate the worth of an investment. The traditional approaches to assessing value are based upon financial theories that suggest that growth in shareholder wealth is a result of the discounted cash flow streams generated by investments.

The time value of money concept is a reflection on the fact that present capital (cash in hand) is more valuable than a similar amount of money received in the future. If this assumption is accepted as fact, then benefits and costs are worth more if they are achieved much earlier. Time value of money computation is based on present value and discounting techniques. The present value method captures the time value of money by adjusting through compounding and discounting cash flows to reflect the increased value of money when invested (Boussabaine and Kirkham 2004).

One of the most important decisions in time value of money is the choice of an appropriate discount rate to be used in the evaluation of PFI/PPP schemes. The choice of a discount rate will affect the acceptance or rejection of investment options under consideration. Hence, the choice of discount is crucial because comparison between options is very sensitive to small changes in the discount rate. Normally, in public project investment, a discount rate is selected to reflect the different costs and benefits that occur at different stages of the investment life cycle. This aspect is usually referred to in literature as the social time preference value of money. It is suggested that for the public sector, projects should use lower discount rates than in the private sector.

In the early phases of PFI/PPP schemes, the UK Treasury uses a real annual discount rate of 6 per cent and a nominal rate of 8 per cent. This is well above the social time value of money discount rate advocated by economics. In the current economical conditions, it is assumed that the social discount rate should be well below the current market discount rate by at least 1 or 2 per cent. Usually these discount rates are chosen based on political dogma rather than economic facts.

The level of the discount rate can increase or decrease depending on tax and inflation considerations. Higher taxation could lead to higher discount rates. But, in practice, the tax effect is disregarded when the cost of capital is estimated. Lower inflation should lead to lower nominal interest rates. The UK Treasury advocates the use of real discount rates in all PFI/PPP project budget estimates.

Capital budgeting principles

Both quantitative and qualitative aspects are taken into account in capital budgeting decisions. Capital budgeting practices are defined as the methods and techniques used to evaluate and select an investment project. These decisions are important because they usually involve large cash outlays with cash flow return over a long period of time. There are four key stages in capital budgeting. These are:

1. Forecast the future project cash flows (both outlays and return) and establish the time when the flow occurs.
2. Estimate the opportunity cost of capital (discount rate). The opportunity cost of capital is the expected rate of return given up by investing in the project under review.
3. Compute the present value of the future cash flows discounted at the opportunity cost of capital rate. The present value of the cash flows represents the maximum amount that investors would pay for the investment.
4. Select the best option based on NPV decision rule.

In PFI/PPP projects, both the public sector and private providers are required to carry out such activity, but based on different assumptions. The public sector carries out their capital budgeting activities to demonstrate affordability and effectiveness of proposed investment. The private providers, on the other hand, perform capital budgeting to evaluate the rate of return they must earn on their capital investment in order to maintain their market value and to continue attracting needed funds for their operations.

In reality, there is a distinct difference between public and private capital budgeting practices. In the public sector, the overriding principle is the provision of services and social welfare based on value of money criteria, whereas the provider's budgeting decisions are mainly concerned with maximising shareholder value or wealth.

The difficulty here is that the public sector must place financial value on both tangible and intangible benefits (see Chapters 2 and 3), and this may lead to difficulties in basing budget decisions on social discount rates rather than market discount rates. Without a doubt, the present value of the benefits from most, if not all, of public projects exceeds costs by a large margin. It is also true that public authorities' responsibility should not be wealth maximisation in a financial sense.

The procurer's budget

This section explains budget development from the purchaser's perspective, at early stages of the PFI/PPP development process. Budget elements by and large include both capital and revenue aspects. A public authority's conceptual budget is normally developed to derive the comparative cost implications of each public investment alternative. Evidence shows that it is more than likely that this initial budget will change substantially. That is why it is important for the client to put in place a robust audit trail to create reference budget assumptions between stages of development. Initial budget as a rule is subject to adjustment upwards, to take into account all possible risks that are associated with the investment proposal. Theoretically, the outcome of such an exercise will form the basis of both financial (i.e., affordability analysis) and economic (i.e., VFM analysis) appraisal of PFI/PPP projects.

A typical cost budget might be based on the following capital assumptions and components:

1 Capital costs

    (a) Asset costs are split into:
        (i) building, 60 years service life and 60 per cent of the overall building cost;
        (ii) engineering services, 25 years service lives and 30 per cent of the overall building cost;
        (iii) equipment/fittings 10 years service lives and 10 per cent of the overall budget (in the majority of PFI schemes equipment is excluded from the deal). Equipment that is usually fixed to structure is included within the building costs.
        (iv) infrastructure and work services.
    (b) professional fees, normally at 15 per cent (or above) of overall capital costs;

(c) Contingencies at 12 per cent of overall capital cost. However, in some schemes the figure is above 25 per cent.
(d) Land, demolition, disposal or purchase costs. However, in PFI schemes the land is usually purchased by the provider.

2 Facilities
Operating costs are split into hard FM and soft FM.
3 Revenues from current services, for example in schools, number of pupils, and hospitals, number of episodes or similar measures. Cash flows are deflated by 2.5 per cent to give real cash flows, which are then discounted at 3.5 per cent for a 30-year period.
4 Residual values of capitals assets used in options selection process.

Based on the above assumption a capital budget is presented in Table 6.1.

Once a project budget has been approved, the client team translates the approval budget, usually related to an option, into a detailed specification output, outcome and desired allocation of risks. Readers can read Chapters 7 and 8 for a more detailed description of costing process.

It is important to stress here that the final budget (a fully costed investment proposal by provider) may well differ quite substantially from the original proposal that has been approved in principle.

The provider's budget

The provider is responsible for the financing, design, construction, repair and maintenance of the building together with the site-wide provision of soft and

Table 6.1 Client outline budget for an NHS scheme

Cost element	*Cost excl. VAT £'000*	*VAT £'000*	*Cost incl. VAT £'000*
1 Department costs based on indicative schedule of accommodation	70000	12250	82250
2 On-costs (60% of 1)	42000	7350	49350
3 Works total (1 + 2) at 174 BCIS 'all in' TPI	112000	19600	131600
4 Location adjustment (at 2% of 3)	3360	588	3948
5 Sub-total (3 + 4)	115360	20188	135548
6 Fees (13% of 5)	14997	2624	17621
7 Non-works costs	n/a	n/a	n/a
8 Equipment costs (at 15% of 5)	17304	3028	230332
9 Contingencies on 5, 6 and 8 (@ 24%)	10519	1840	12359
10 Total	158181	27681	185863
11 Inflation adjustment	n/a	n/a	n/a
12 Project case total	158181	27681	185863

Note: Price base = 2005.

hard FM services. The project company is responsible for all aspects of the PFI/PPP works and delivery of the specified contract services throughout the operational period of the contract. The provider may also enter into sub-contracts with specialist companies for the provision of the non-specialised services.

Most of the information in the provider's budget is commercially sensitive. However, the following can provide some idea of the general aspects of the pre-operation budget of the project company:

- Maximum funding requirement:
  - equity 9 per cent of the total budget;
  - senior debt 91 per cent of the total budget.
- Pre-operation capital expenditure:
  - design and construction costs 75 per cent of the total funding;
  - set-up costs 11 per cent of the total funding;
  - construction period operating costs 2 per cent of the total funding;
  - interest during construction 6 per cent of the total funding;
  - debt fees 2 per cent of the total funding;
  - pre-funding of reserves 4 per cent of the total funding.
- Operation expenditure (for an assumed acute health scheme):
  - hard FM at 11 per cent of the payment waterfall (assumed here as unitary payment);
  - soft FM at 20 per cent of the payment;
  - mechanical and electrical service (MES) at 7 per cent of the payment;
  - central sterile services (CSS) at 2 per cent of the payment;
  - senior debt at 40 per cent of the payment;
  - SPV operating costs at 20 per cent of the payment.
- Revenue or waterfall:
  - unitary payment (assuming no other incomes).

Based on the above assumptions, an initial capital budget for a project company is presented in Table 6.2.

The contractors' budget

Generally, construction contractors are responsible for the design and construction of PFI/PPP schemes. This is a fixed lump sum contract.

A typical construction budget in most cases is based on the following capital assumptions and components:

Table 6.2 The provider's initial budget

Funding			*Construction all inclusive*			*Operation revenue (after the construction period) all inclusive*		
Source	%	*£'000*	*Uses of funds*	%	*£'000*	*Uses of payment*	%	*£'000*
Equity	9	9000	Design and build costs	75	75000	Hard FM	11	2750
Senior debt	91	91000	Setup costs	11	1000	Soft FM	20	5000
			Construction period operation costs	2	2000	MES	7	1750
			Interest during construction	6	6000	CSS	2	500
			Debt fees	2	2000	senior debt	40	10000
			Pre-funding of reserves	4	4000	SPV operating costs	20	5000

Notes: Total project funding = £100,000,000. Unitary payment = £25,000.

- Construction:
  - substructure/foundation at 2 per cent of the total construction cost;
  - superstructure at 22 per cent of the total construction cost;
  - roofing at 2 per cent of the total construction cost;
  - mechanical at 30 per cent of the total construction cost;
  - electrical at 10 per cent of the total construction cost;
  - exterior closer at 10 per cent of the total construction cost;
  - interior closer at 19 per cent of the total construction cost;
  - exterior works at 5 per cent of the total construction cost.
- Professional fees (including design) usually at 15 per cent of the total project cost excluding operational costs.
- Contingencies at 12 per cent or over of the total construction cost.

This breakdown serves as an indication of how a contractor budget is split between different cost elements. One should expect a slight variation in the percentages depending on the type of project under estimation. Contractors' construction budgets have been the subject of many investigations and readers can consult specialised text books for further details.

Measuring the risk in time value of money decisions

Risk exists in time value of money decisions because each input time value parameter has a number of possible outcomes, which in turn relate risk to an uncertainty outcome. Risk in time value decisions results from uncertainty, risk and uncertainty hypothetically are not identical (see Chapters 12 and 13).

Risks in time value decisions are normally associated with decision-making situations where the probability of time value parameters, such as PV, interest rate, service life parameters and outcome are known, while uncertainty exists when the probability is not known (Lefley 1997). Choobineh and Behrens (1992) support the view that risk is the consequence of taking a time value decision in the presence of uncertainty, while uncertainty is the manifestation of unknown consequences of change in time value parameters. Based on this paradigm, one can define the uncertainty in time value as the gap between the information currently available and the information required to make the decision on these issues. A condition of uncertainty usually exists in time value parameters because investment decisions, by definition, involve uncertain outcomes that in the long run are important to project stakeholders' survival and about which complete information is unavailable.

One may argue that risk in time value parameters increases with time; the longer the project concession agreement, the greater the difficulty in estimating these parameters, especially towards the end of the concession. Hence, the shorter the period of analysis, the lower the risk, while study over longer periods, which is the case in PFI/PPP projects, indicates a higher risk in achieving the forecast values.

There are a number of methods used in the industry to identify and assess the level of perceived time value risks. The best-known and used methods are detailed in the following section.

Adjustment of discount rate

A survey by Lefley and Sarkis (1997) found that a considerable number of companies adjust discount rates in order to take risks into account in projects' appraisal. Others (Allen 1992) argue that it is better not to adjust the discount rate; instead, it is good practice to use a riskless discount rate and only adjust the projected cash flows to take into account the anticipated risks. The question that needs to be answered here is whether time and risks in discount rate evaluation are separate variables. If the answer is yes, then one could ask, 'Does adjusted discount rate depend on time or risks?' Probably on both; if this is the case, then probably it will be very difficult to establish a risk-free discount rate of return that includes an adjustment for investment risk. This view is supported by Aggarwal (1980) who stated that it is complex, if not impossible, to precisely establish a discount rate that will reflect the level of risk associated with investment in projects.

Probabilistic methods

Various time value investments parameters cannot be determined with any degree of certainty. The value of each of these parameters might be affected by a myriad of risks and uncertainties which are not easy to quantify over a long-term horizon such as in PFI/PPP projects. Probability distribution of possible returns is frequently used to measure risks in investments. The dispersion of the probability of possible returns reflects the degree of uncertainty that is present in PFI/PPP projects. A distribution with small standard deviation relative to its expected value indicates little dispersion and high confidence in the outcome. In contrast, a distribution with large standard deviation relative to its expected value indicates a high degree of uncertainty about the possible return on investments. For further details on probability distributions and associated parameter readers, consult Boussabaine and Kirkham (2004).

Sensitivity analysis

The use of a series of discount rates to test the sensitivity of project outcomes to different rates is also advocated by many economists. Sensitivity analysis

> measures the impact on project outcomes of changing one or more key inputs values (assumptions) about which there is uncertainty . . . sensitivity analysis is normally performed altering one assumption at a time, but it is possible to adjust more than one assumption simultaneously.

Further details are provided in Chapter 12.

It is imperative that PFI/PPP stakeholders should seek a consensus approach to decision-making in relation to uncertainty and risk in selecting financial investment parameters.

Economic evaluation of projects

Capital budgeting methods, capital budgeting techniques and project evaluation techniques in most of the literature have identical meanings and are normally used alternatively. Studies show that the most widely used project evolution techniques are:

1 payback method
2 net present value
3 internal rate of return
4 accounting ratios
5 annuity
6 value-based management
7 real option.

These techniques help investment decision-makers to select n out of N investment projects with the highest profits at an acceptable risk of loss. To determine the most appropriate project economic evaluation method, organisations should consider the net benefits of capital budgeting methods and tools to their costs. It is also normal practice to combine different economic evaluation methods to provide additional information on taking investment decisions.

The most widely used project evaluations in the construction industry are payback period (PBP), net present value (NPV), internal rate of return (IRR) and to some extent accounting ratios in PFI/PPP projects (see Chapter 5). This could be attributed to the fact that the private sector shareholders demand value added from their investment. The easy way of demonstrating value in pre-investment stages is by showing that an increase in shareholders' value is expected when a positive NPV occurs. Hence, in this section we will only discuss PBM, NPV and IRR.

Payback method (PBM)

This is a simple measure, easy to use, understand, interpret and implement. However, the payback method does not show value enhancing results. The payback method period is related to liquidity rather then an increase in shareholder value.

It can be looked at as a cash flow method, if the cash flows are adjusted with the cost of capital (i.e., discounted payback period).

The most important decision criterion for the payback method is the number of years required for the recovery of all payments. It is expected that payback periods in PFI/PPP projects are determined on a project-by-project basis. However, the following general rules should be considered:

1 Use high hurdle rates and long payback periods for large or strategic investments.
2 Use low hurdle rates and short payback periods for small investments.

The drawbacks of the PBM method are:

- It does not consider the time value of money.
- It does not incorporate risks in systematic way.
- It does not use cash flows.
- It does not account for all cash flows occurring after the payback period.
- It gives equal weight to all cash flows arriving before the cut-off period.
- The timing of the cash flows and the cash flows beyond the payback period are not considered.
- There is bias against longer-term projects.
- It can lead to the rejection of positive NPV projects.

PBP decision rules:

1 A project is acceptable if its payback is less than the maximum cost recovery time established by the analyst.
2 The investment should proceed if the payback period exceeds a specified period.
3 When using PBP as a ranking method between projects, the project with the shortest payback period should be selected.

Net present value (NPV)

The net present value method computes the difference between values of the investment from the present value of the benefits of a project. The benefit should include both tangible (revenue) and intangible benefits.

It is one of the most widely used methods because of its ability to consider all cash flows of projects as well as accounting for the pattern and timing of the cash flows. An NPV above zero is expected to guarantee that the value of the consequences is at least as high as the value of the resources consumed. The same is true for public sector services, which are expected to demonstrate that the value from the provided services is at least as high as the value of the capital resources consumed.

NPV disadvantages include:

- It requires and assumes equal risks for both cash inflows and outflows.
- It may lead to different decision outcomes.
- Mutually exclusive projects assume that investment opportunities are independent of each other. In reality, most of the investments are interdependent.
- It takes no consideration of the new debt required to refinance projects if needed.
- It does not take into account opportunity cost.
- It favours options that defer expenditure over those that have high costs in early years.
- Difficulty of forecasting cash flows from long-term projects, which casts doubt over the appropriateness of the NPV rule in capital budgeting especially in PFI/PPP contracts with a concession period of over 25 years.

NPV advantages are:

- It considers the forecast future cash flows of the investment.
- It considers the timing of the cash flows–present value analysis.
- It considers the opportunity rate of return of shareholders, the project's riskiness relative to others, and alternative investor opportunities.
- It considers the time value of money.

NPV decision rules are based on the paradigm that investment opportunities instantly disappear if they are not immediately undertaken. The general rules are summarised as follows:

1 An NPV much larger then zero is required to exercise the option to invest.
2 If the net present value is negative, because the return (the appropriate market rate) is higher than the return from this project, the opportunity value is negative, then decline the investment.
3 If the present value is the maximum amount one would pay for the investment relative to the cost, the NPV represents the added value of the investment.
4 Using NPV as a ranking method between projects, the project with the largest positive NPV is selected (in the case of the public sector, the option with the lowest NPV is selected).
5 If the NPV is equal to zero, then the decision to invest or not is indifferent.
6 If there are many projects, mutually exclusive, and there is no budget constraint, then rank by positive NPV > 0 and select the one with the largest NPV, since this project maximises the size of the return.

Internal rate of return (IRR)

In most literature the rate of return, the internal rate of return (IRR), or the discounted cash flow (DCF) rate of return have identical meanings and are normally used alternatively. IRR is the interest rate which makes the NPV of the project zero. IRR provides the rate of return that an investment will provide if it is accepted. Normally this is reported as a percentage benefit from the given investment. Theoretically, IRR is the rate that equates the cost and benefits of the project in terms of present value. So at the rate of IRR, the NPV of the investment will be zero. This implies that IRR is the maximum cost of financing the investment.

The drawbacks of using IRR are:

- It assumes the same rate of lending and borrowing.
- It is not suitable for non-conventional patterns of project cash flows.
- It takes no consideration of the new debt required to refinance projects if needed.
- It does not take into account opportunity cost.
- Values are computed before investment occurs. Changes to computation assumptions during the project operation phase will be very difficult to rectify.
- Any change in cash flows' timing produces as many IRRs as there are changes in the cash flow directions of the investment.
- Limited capacity to identify the value creating/destroying elements in a project.
- Small projects may be erroneously selected over larger projects using the IRR (mutually exclusive projects' conditions).
- Different time profiles of costs and benefits may result in ambiguous ranking.

IRR advantages are:

- It considers the forecast future cash flows of the investment.
- It considers the timing of the cash flows–present value analysis.
- It considers the opportunity rate of return of shareholders, the project riskiness relative to others, and alternative investor opportunities.

IRR decision rules are:

1. Accept projects where the IRR is greater than the opportunity cost of capital (discount rate).
2. Reject projects where the IRR is less than the opportunity cost of capital.
3. Using IRR as a ranking method between independent projects, the project with the higher IRR is preferred (there may be exceptions to this rule).
4. If projects are mutually exclusive, we cannot rank them by their internal rates of return. This is because the IRR is independent of the size or scale of the project in terms of time.

The financial criteria for PFI/PPP projects' appraisal

From the public sector point of view, projects' economic appraisal depends upon two financial criteria: value of money (VFM) and affordability. In the UK, it is a requirement that public authorities must demonstrate through the use of discounted cash flow methods (mainly NPV) that the private procurement option is financially better than the public sector. This process is based on what is known as a public sector comparator (PSC); readers can consult Chapters 2 and 3 for the theoretical background of VFM and value creation. Affordability is established through projection of income from the services provided over the life span of the concession. This must include both tangible and intangible benefits. As stated above, in a commercial setting, the NPV investment criteria state that the investment with the highest NPV is preferred because it maximises wealth to shareholders. However, in PFI/PPP investment appraisals, from a public point of view, the option with the lowest net present cost is preferred as yielding the greatest financial benefit. It is also expected that the costs of the risks associated with projects are included in NPV computation. The problem here is that it is not possible to equate public investment with the private sector. The public sector investment is service- and welfare-driven rather than market opportunity-driven. Also, deduction from NPV results is not appropriate when capital is rationed, which is the case in most public sector projects; that is why finance theory does not consider or take into account NPV results under such conditions.

Using cost indices in PFI/PPP projects

The main purpose of indices is to capture changes in the cost of building elements, inflation, regional price variations, contract conditions and service payments from one point in time to another. There are several types of indices that might be used in PFI/PPP projects cost planning. This might include:

- retail prices index;
- median index of public sector building tender prices;
- index of basic materials cost;
- index of wages or labour;
- general building price index;
- index of specific building types and works;
- tender price index.

Since the period between the selection of a preferred bidder and financial close could span over several months, all construction prices are adjusted to inflation using MPIS indexation (Median Index of Public Sector Building Tender Prices produced by the UK Government for the public sector, calculated from rates for work and bills of quantities from accepted tenders). This is also true for outlined business cases that are prepared by purchaser authorities. Hence, construction cost inflation is normally taken into consideration by applying the construction cost indexation rates specified in the project agreement (part of the inputs into the financial model). Where the construction cost breakdown is not clearly defined in the financial model, the cost is usually calculated by reference to a fair valuation for the cost of complying with the variation.

In the pre-financial close, this covers the increase in the FTN prices during the period from the date of the estimate to the expected financial close, or other specified dates, i.e., when the increase is added to the estimated current building cost, the total will equal the anticipated tender amount. The following formula can be used to compute the increase in inflation:

$$Y = Ec^{*}Eai^{*}Mn$$

Where:

Ec is the estimated current building cost.

Eai is the estimated average monthly percentage increase.

Mn is the number of months from the date of the estimate to the tender date.

Also the following formula can be used:

$$Y = (MIPSc -)\ MIPSb/MIPSb^{*}\ Ec + Ec$$

In PFI/PPP projects that are based on service payment mechanisms, if the service payment is set for the project term, the providers usually seek to protect themselves against the effects of inflation through appropriate adjustments to the service payments. To achieve this, the service charge is usually indexed over the project term. Hence, payment modalities in PFI/PPP schemes are subject to many adjustments to take into account variations, market testing, changes in law, and as a result of insurance premium. The adjustment is measured by changes in the relevant index published for that contract year and may be calculated in accordance with the following formula:

Amount or sum in (date) prices × RPId/RPI0

Where:

RPId = RPI in the current year or the value of the retail prices index published or determined with respect to a specific period as stated in a project agreement.

RPI0 = the previous year or the value of the RPI in respect of a date.

Project cash flow

Under PFI/PPP schemes both clients and provider are required to forecast cash flows for at least the concession period. Harris and McCaffer (1989) state: the factors that affect project cash flows include:

- duration;
- profit margin;
- the retention conditions;
- delays in receiving payments from either SPV and lenders;
- credit arrangement with suppliers;
- settlement of outstanding claims.

It is argued that the uncertainty of forecasting future cash flows increases with time; the longer the concession, the greater the difficulty in estimating cash flows, particularly towards the end of the concession.

Cash flow at the construction stage

All pre-financial close costs are included as part of the project's cost. Preferred bidder costs are capitalised. Many PFI contracts allow for recovery of bid costs at financial close. All construction phase costs, including the operating costs of the SPV, are drawn down from debt and equity. Construction and financing costs are charged to PFI assets in the construction phase. Funding to cover this is drawn against non-recourse debt-financing PFI assets. But in some cases equity debt is spent first. During the construction period, the SPV is normally invoiced by construction contractors and subcontractors for construction revenue. The procedure and the amount of drawing construction funds usually are defined in the project agreement and are also based on the construction progress. Figure 6.1 shows a typical cash flow curve of drawdown payments.

Contractors and subcontractors should expect to experience negative cash flows in the early stages of the project. Construction contractors' payments are normally made on a milestone or completed work basis, with retention as shown in Figure 6.2. Construction is normally procured under a fixed price turnkey design and build contract.

SPV cash flow

Under all PFI/PPP projects, the SPV experiences negative cash flow until the start of the operational stage; this phenomenon represents locked capital that is supplied from senior debt. In return for this, the project company will be remunerated, once the asset is available for use, by receipt of regular payments, and these will be indexed to inflation over the concession period to cover service of capital together with operating, maintenance and ancillary service costs.

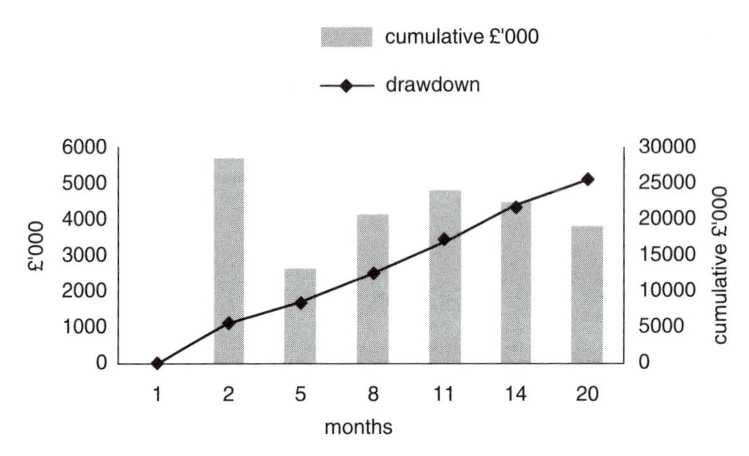

Figure 6.1 A typical debt drawdown cash flow profile for a 20-month construction contract.

Hence, the SPV cash inflows only start after assets and services are up and running. Project company cash inflows are, in most cases, based on either the criterion of availability (mainly in schools, hospitals, prisons, etc.) or throughput (e.g., in roads and power stations). The SPV's cash inflows are based on a series of payments over the life of the concession (by and large invariable in real prices but inflated over time by RPI) based on the extent to which the buildings are available for use as defined in the project agreement. Where availability is less than contractually anticipated, a payment deduction is applied according to

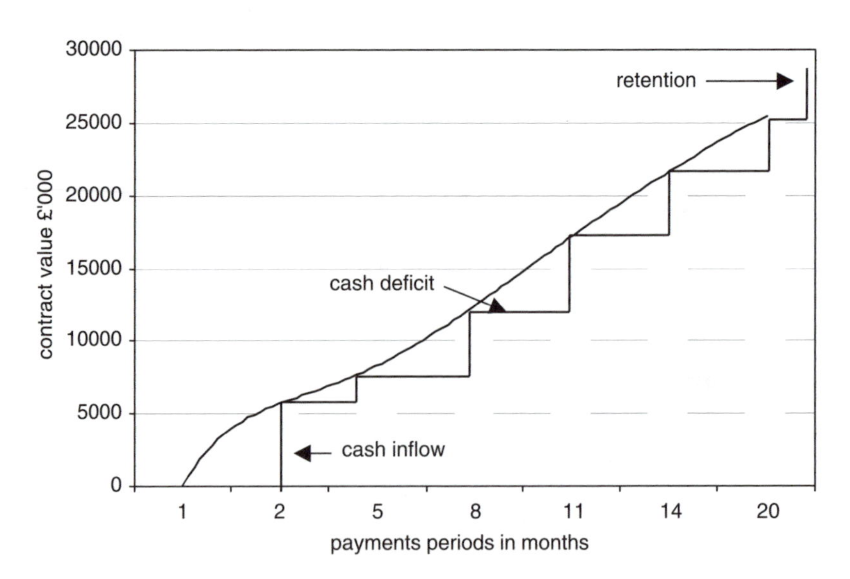

Figure 6.2 A typical contractor cash flow profile.

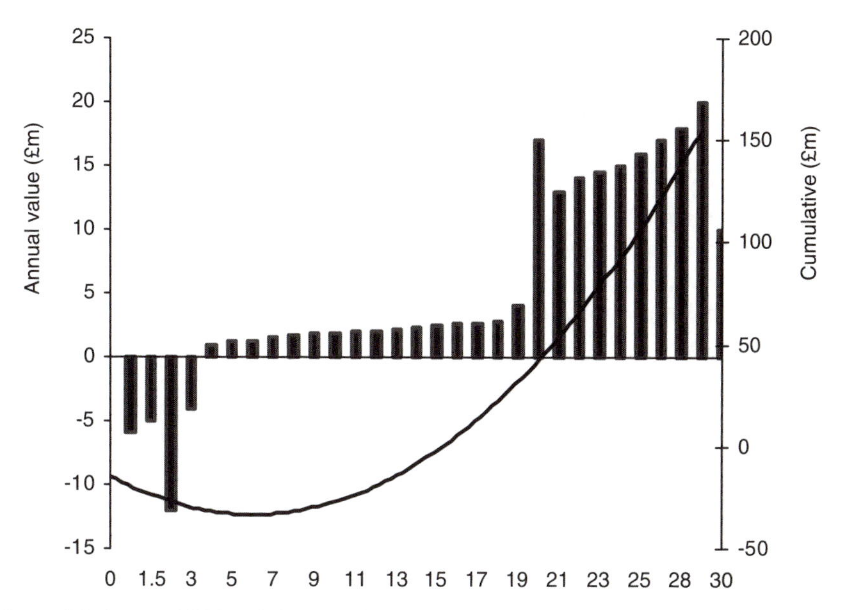

Figure 6.3 A typical SPV cash flow profile.

contractual formulae defined in the project agreement. Based on these cash inflows paradigms, the project company ought to forecast future cash projections by correlating required availability of buildings with known performance by simulating all possible scenarios. Even with this in mind, loss of income could occur because of subcontractors' inability to deliver the required performance. In such scenarios, the SPV obtains damages to recover the shortfall in income due to the subcontractors' default. Figure 6.3 shows an example of the cash outflows of a typical SPV based on a generic project with service commencement after three years of construction.

The majority of the cashflow in the first three years of the concession relates to the construction of the asset. The SPV is allowed to draw down payments to fund work in progress as well as funding the costs of running the project company. Generally, this is done on an agreed schedule in the project agreement. It is important to point out here that there are no profits earned in the construction period. As demonstrated in Figure 6.3, after the construction period, the SPV starts to receive operational payments for the use of the facility and revenue increases to reflect the operation and maintenance of the asset over the remainder of the concession period. The SPV operates in cash deficit for most of the project concession period. This is due to the depreciation charge on the constructed asset combined with the interest charge on the outstanding loans. Ultimately the SPV's profit will increase when the interest charge on the outstanding loans falls. The SPV will be in a position of cash surplus towards the beginning of the third period of the concession duration. From this point on, the SPV's profits rise progressively to the end of the concession, reflecting the reduced interest charge as the loans are repaid.

The SPV cash outflows at the operation stage are normally based on the following expenditures (Balfour Beatty 2005):

- *Operating costs*: payment of the operations and maintenance subcontractors.
- *SPV operating costs*: these include personnel costs, office and equipment, training, etc.
- *Insurance*: this includes insurance against all risks to cover the project against physical damage.
- *Debt service*: interest and principal repayments on the debt funding for the project. The interest payments reduce over time as the principal is reduced.
- *Tax*: due to capital allowances and other deductions incurred in the construction phase, tax does not become payable by the project company until typically around 10–15 years. Thereafter, tax is payable on the profits earned.

Cash flow waterfall

The cash earned by the SPV from its revenues is distributed according to strict rules in the PA which set out the order of priorities for the use of this cash.

SPV revenue from payments is normally spent in the following order:

- funding operational services (working capital) including the running of the SPV and taxes;
- interest payments, i.e., interest on the debt and hedging payments;
- amortisation, i.e., debt repayments according to a specified schedule;
- funding of debt service reserves to provide security against short-term cash flow problems;
- funding of life cycle replacement reserves.

Any cash left from the above commitment is the free cash available to shareholders of the project company. This is distributed by way of dividend and subordinated debt interest payments under the following conditions (subject to specified minimum debt service coverage ratios):

- Equity providers receive dividend income towards the end of the project concession.
- Subordinate debt holders start to receive payments after the completion of the construction phase.
- Towards the end of the concession, any free cash held as cover for reserve accounts is distributed.

In most PFI/PPP contracts the providers use refinancing to reduce the total cost of debt. The reason for this is that the initial financing rates account for the construction risk being higher than the enduring risk. However, following the construction phase, it is sometimes possible to achieve a better rate due to the reduced risks in the operating contract phase. That is why most providers use this risk reduction paradigm, to obtain cash gains from refinancing, as part of their investment strategy in PFI/PPP projects.

Summary

The time value-based investment concepts are employed to evaluate the PFI/PPP schemes for value for money aspects. There are a number of methods used to identify and assess the level of perceived time value risks. These include adjustment of the discount rate, probabilistic methods and sensitivity analysis. From the public sector point of view, a project's economic appraisal depends upon two financial criteria: value of VFM and affordability. A project company's cash flow rises progressively to the end of the concession, reflecting the reduced interest charge as the loans are repaid. The project company will be in cash surplus towards the beginning of the third period of the concession duration.

Chapter 7

The process of cost planning PFI/PPP projects

Introduction

Cost planning of PFI/PPP projects must provide an exhaustive accounting of all resources required for the design, construction operation, maintenance and disposal of assets. It is well recognised that the majority of decisions about these costs are predetermined at an early stage of a project's development. The opportunities to modify or influence costing decisions decrease as projects progress through their life cycle. That is why it is important to establish a robust mechanism at the early stages of development that brings together all the components of cost planning from the perspective of both the clients and providers. To achieve this aim, the cost planning of these projects requires a systematic framework that considers all aspects of projects' development and operation. A framework that permits cost data flow between cost planning stages and that provides benchmarks and guidance within which cost control may be exercised is essential for the effective costing of PFI/PPP deals. This chapter will deal with PFI/PPP cost planning processes and methods for managing such procedures.

The costing process

The aim of cost planning is to instigate a risk reduction strategy. This will ensure that project development budget threats are identified and assessed accordingly.

The process of cost planning in PFI/PPP projects procurement has not been documented at all. This is mainly due to sensitive commercial confidentiality within this procurement system. The process is mainly driven by legally defined documentation such as the project agreement (see Chapter 6). This section attempts to define and lay down the steps involved in the cost planning of PFI/PPP projects from both the clients' and the providers' point of view.

Procurer's cost planning process

The procurer's planning process can be summarised in Figure 7.1. Estimates are prepared at various stages of the cost planning process. In the preconstruction

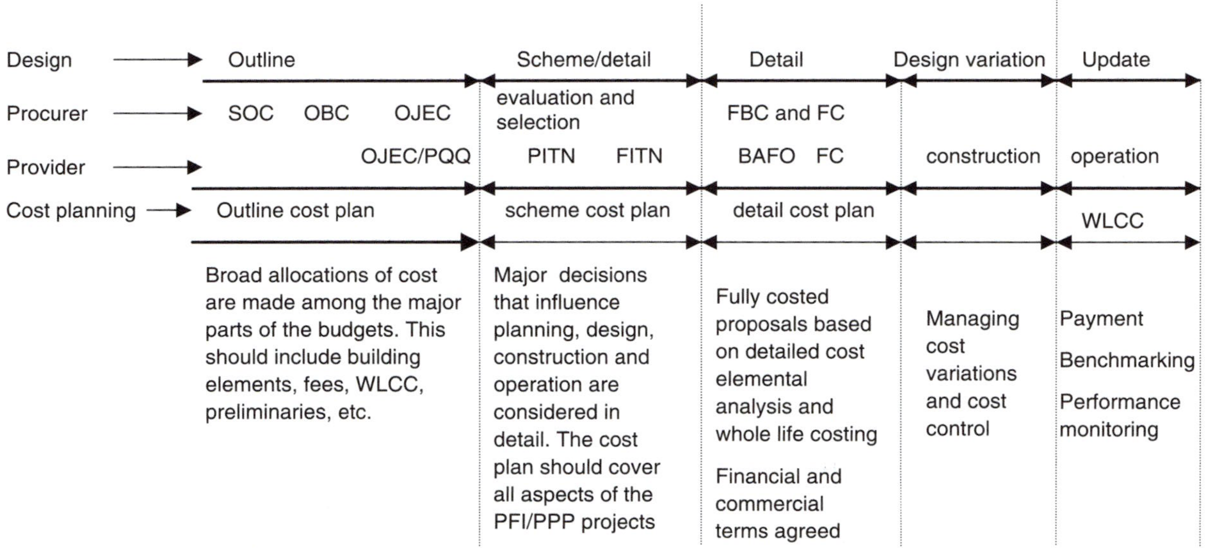

Design	Outline	Scheme/detail	Detail	Design variation	Update
Procurer	SOC OBC OJEC	evaluation and selection	FBC and FC		
Provider	OJEC/PQQ	PITN FITN	BAFO FC	construction	operation
Cost planning	Outline cost plan	scheme cost plan	detail cost plan		WLCC
	Broad allocations of cost are made among the major parts of the budgets. This should include building elements, fees, WLCC, preliminaries, etc.	Major decisions that influence planning, design, construction and operation are considered in detail. The cost plan should cover all aspects of the PFI/PPP projects	Fully costed proposals based on detailed cost elemental analysis and whole life costing Financial and commercial terms agreed	Managing cost variations and cost control	Payment Benchmarking Performance monitoring

Figure 7.1 Design, procurement and cost planning processes.

stage, to assist in the development of the cost plan, the budget is normally analysed into capital and financial appraisals.

During the early stage of PFI/PPP projects' development, general preliminary estimates are usually carried out. Public sector comparator, capital appraisal, financial analysis and affordability studies are based on such estimates. It is suggested that in the absence of complete design and specifications, preliminary cost plans and estimates that are lacking a rational basis may prove disastrous for project stakeholders (Li *et al.* 2005). It is well documented that the error of cost estimates at early stages of projects' development can be over 20 per cent of the final cost. This may have a significant effect on using public sector comparators. An increase in the capital cost may make the public sector option the preferred one that delivers value for money.

Design drawings and a schedule of accommodation are developed from the service output specifications (SOS). Also, in the early stages, cost plans are based on the schedule of accommodation. As shown in Figure 7.1, for each phase of the PFI/PPP procurement process, a cost plan is produced in a progressive manner. The progression between stages is subject to signing off by appropriate authorities (in most cases the project board). It is important in cost planning processes to have key sign-offs to consolidate the previous stage costing and to trigger the next loop of the iterative process of cost planning. This means that the budget at certain key stages will be frozen with the exception of the issues that have been designated by the bidders for further discussion.

Cost planning at the SOC stage

Project initiation (SOC phase) covers the initial stages of the project cost planning based on general information on services requirements, equipment, design brief, etc. Requirements for future services are documented by the public procurer departments. This documentation is then used to help determine the accommodation and services required for PFI/PPP schemes. The output from this exercise is a list of schedules of accommodation. The accommodation requirements are then costed and the total expenditure required is projected based on feasibility studies and the design brief. Hence, the SOC appraises the investment options in terms of both service and financial benefits. The SOC sets out why the development is needed, examines options and puts forward a case for investment. Theoretically, the SOC does not decide whether the scheme should be financed via PFI but only whether it should go forward for assessment. At this stage, in most cases, cost plans from the gross internal area of service support space (e.g., clinical and non-clinical) are required for each component of each option under examination.

Capital costs are normally developed from the gross internal area (or similar methods) of accommodation space required for each component of each option. Whole life costs are also forecast for each short-listed option. The contingency associated with planning, construction and commissioning is usually estimated at this stage as a percentage of capital costs. An example of this early cost planning is shown in Table 6.1. It is imperative to state here that improbable costing at this stage may lead to unrealistic comparison between options.

Cost planning at the OBC stage

If the SOC is approved by the appropriate authority concerned, the procurer will be asked to draw up an outline business case (OBC). This will examine in detail whether the project should be financed and provided by the PFI. It does this by comparing the PFI option with a theoretical public sector comparator (PSC). The costing details from the SOC stage will feed into the OBC stage. Hence, the second gateway (OBC) covers a more detailed cost planning process that includes capital and financial appraisals. This should also include the pricing of service output specifications, option appraisal, and public sector comparator. The cost plans developed at earlier stages should now be updated to reflect more detailed design information that has emerged from the outline design. The purpose of cost planning here is to confirm whether the PFI/PPP procurement route is the preferred option. The OJEU/PQQ gateway stage starts with the approval of the PFI/PPP procurement route. The aim of the PQQ process is to measure the capacity and capability of bidders (provider candidates) to undertake the project in question. PQQ provides the procurer with the opportunity to improve and update their OBC cost plans. This refined OBC will then become the basis on which PITN will be based. At this stage, the OBC cost plans will be further developed to prepare PITN documentations. The PITN gateway phase provides a framework for the pre-qualified bidders to put together their proposals and to demonstrate their ability to bring innovation and value for money to the scheme under consideration. Cost planning here is based on an outline design scheme that indicates the functional relationships and area schedules. Outline architectural drawings are also used to price the schemes at this stage. Cost planning here consists of economic appraisal to assess the value for money for the proposed scheme. An economic appraisal is normally carried out to consider which of the options represents the best value for money to the client. Discounted cash flow analysis, using the net present value methods, is used to compare the options. Value for money is measured through appraisal to determine and compare the NPV of the PFI option and the PSC. In most cases it is a requirement that cash flows should exclude capital charges and VAT. It is imperative that the appraisal must include adjustments to reflect the estimated monetary value of the risks retained by the client to measure the allocation of risks under each option. The ultimate aim of the analysis is to arrive at a preferred option. Once this is done, then the option is developed into a reference project, comprising the services and the associated facilities to be procured. It is costed in details to indicate the likely cost of the PFI/PPP option.

It is important to stress here that depending on a particular project's complexity and value, the client may waive the requirement for a PITN/ISOP submission.

The FITN gateway stage solicits priced bids from two or three short-listed bidders (in some cases, up to five bidders are short-listed). It is normal at this stage to include a draft contract agreement, payment mechanism and performance measurement system in the FITN documents. Following the issue of the FITN documentation, the procurer's team meet each bidder to develop the scheme layout details. Also issues relating to finance, legal, human resources and facilities management are discussed and specified in some detail. OBC cost planning at this stage includes detailed elemental cost analysis provided by bidders in response to the FITN.

Prior to proceeding to the BAFO stage, bidders may be asked to provide written confirmation of the fixed price that they submitted in response to the last stage. It is also common practice that bidders include in their FITN cost plans submission issues that are open to further negotiations, if they are being selected as preferred bidders. Normally, two bidders will be taken forward to the BAFO stage based on the fact that their bids may be potentially affordable and their service specifications are acceptable as set in the selection criteria for a particular scheme (in some cases the client may proceed straight to selection of a preferred bidder). After the selection of a preferred bidder, the design development process will continue up to financial close and beyond to the construction stage, as shown in Figure 7.1. During the BAFO stage, meetings are normally held between the design teams from each of the bidders and the purchaser user groups. This is essential for all the contracting parties to develop cost and design solutions that meet the procurer's service requirements. The bidders are usually asked to meet all bid requirements set out in the BAFO documentations. It is speculated that at this stage preferred bidders may downgrade the specification to meet client budget and in particular to meet the affordability criteria. This could be due to the fact that unrealistic cost plans have been developed in the previous versions of the OBC. Cost plans here are developed based on a detailed design and full services specification. Cost planning here involves all key elements of the PFI option (design and construction, facilities services and project finance). These cost plans are subject to rigorous appraisal by the client and their expert advisers.

Cost planning at the FBC stage

The investment appraisal undertaken in the course of the BAFO bid evaluation will form the basis of the PFI/PPP Full Business Case (FBC) which will update and re-address all the issues contained in the OBC prepared at previous stages. The FBC will probably include price certainty.

Procurers normally at the financial close (FC) phase enter into the final negotiations stage with the preferred provider leading to completion of the contract and the FBC. Cost planning here involves consolidating the budget and checking that the overall cost plan remains within the affordability and PSC assumptions. Costs provided by suppliers in the course of the procurement will confirm if the cost plans developed in the SOC, OBC and FC stages are accurate.

Provider's cost planning process

The provider's cost planning and bidding processes for PPP/PFI projects vary depending on the particular market sector, the public sector authority involved and they change from project to project. As shown in Figure 7.1, the provider involvement in PFI/PPP schemes start at the OJEU notice. However, many providers select the project types, locations and clients that they want to get involved with well before the OJEU stage. Normally a team is assembled prior to this stage to deal with the bidding cost planning and bidding processes.

Cost planning at the PITN stage

The PITN documentation represents the basis on which bidders develop their cost plans. Their costing is based on output specification which includes outline requirements relating to design and facilities management. In some sectors the PITN stage is by-passed. The PITN stage submission requires bidders to develop outline proposals for the specific project, which may include outline design proposals and preliminary cost plans. Normally bidders use the PITN supporting documentation (produced by the client) to develop their cost plans. These may include the following:

- brief
- output specification
- room data sheets
- accommodation schedules/adjacencies
- title deeds and boundary plans
- topographical surveys
- ground investigation reports
- asbestos surveys
- condition surveys.

In general, the PITN documentation includes an overall sketch of design layout plans. This, along with other information, is used by tenders to develop preliminary cost plans. According to the design development for PFI schemes, the bidders' PITN cost plans should consist of the following (DH 2004):

- functional content
- floor area
- capital cost

- equipment
- external works and site infrastructure
- fees, overheads and profit
- facility management
- whole life cycle costs.

As can be seen from the above, the outcome of a PITN cost plan is a broad cost allocation for the major components of the scheme. It is important that the client's OBC is understood so that operational costs and capital costs are captured.

Cost planning at the FITN stage

At the FITN stage, providers are required to develop detailed proposals covering all aspects including design, construction, operation, management, whole life cost, insurance and finance of the project. Detailed cost and financial models are constructed and optimised by each bidder and its advisers to ensure that the most economically advantageous bid is submitted to conform to the stipulation laid down in the tender documentation by the client. The cost plan at this stage should follow the design progress. By the end of this stage the design will be more or less in an advanced stage. Hence, detailed elemental measurement and unit rates are required to estimate the capital and whole life costs. Cost plans are based on a detailed elemental breakdown of the project. Hence, the provider's FITN cost plans may fall into two key parts. The financial structure and the deliverability of the funding package. The cost plans should include full details on the proposed funding structure, setting out clearly the institutions involved, the type of facility to be provided, the amount of funds available and the terms on which these funds are being offered. Also required at this stage are the support and confirmation from funders that they have reviewed and accepted the project agreement, the payment mechanism, the financial model outputs and that the financial model accurately reflects the lender's terms and conditions where margins, fees, cover ratios, and other covenants and tenors are concerned (see Chapter 8 for further details).

The fully costed bid proposals

Providers are expected to give financial projections for the full duration of the contract period (starting from the financial close date) in the form of a dynamic financial model (capable of being used to run sensitivities in the key areas of risk that the lenders and rating agencies are likely to focus on). The model in most cases includes the following requirements:

- a series of financial schedules (e.g., projected profit and loss account);
- a price base date;

- economic assumptions (e.g., discount rate, inflation rate);
- cost allocation assumptions (e.g., bid development costs, design on costs, fixed SPV running costs, etc.);
- taxation and accounting assumptions;
- funding costs assumptions (e.g., funders' margin, interest rate, etc.).

Detailed breakdown of capital costs

In general, capital costs are part of the bidders' deliverables (fully costed bid). The provider's capital cost plan should at least cover the following major cost elements:

- design and construction costs;
- development (set-up) costs analysis;
- life cycle costs, broken down into the categories;
- facilities management services expenditure – split between fixed and variable costs, where applicable, and including details for each service;
- maintenance and debt service reserve accounts in accordance with lenders' requirements;
- reserve accounts required by the lenders;
- equipment costs.

In most projects, it is a requirement that providers are required to supply a detailed and fully quantified cost plan, which arithmetically supports the costs associated with these major capital cost breakdown elements. The cost plan must be presented in sufficient detail to identify the specification, quantity and rate for all major materials and components adopted for the builder's work, the engineering installations, and the whole life cycle costs, including operational costs. Separate elemental breakdowns are usually developed for each part of the construction works. Chapter 8 provides a detailed account of capital cost plan development.

Cost planning at the BAFO stage

A BAFO stage is not required in all cases and it is stated in the ITN documentation if it is to be used. Bidders normally use the BAFO phase to negotiate and refine the key commercial and financial terms. Fully costed proposals are confirmed at this stage. Cost planning here will include the same elements as in the previous stage, but with cost certainty, as well as other financial and commercial plans. Confirmation and sensitivity of the following financial terms are required at this stage:

- interest rate changes;
- inflation/indexation changes;

- corporation tax and VAT changes;
- construction cost increases;
- life cycle cost variations;
- operating cost variations;
- payment deductions;
- delays in achieving financial close;
- construction slippage in the programme;
- early completion;
- revenue variations;
- drawdown schedule;
- repayment schedule and tenor.

Collecting the relevant cost information

Data can be obtained from a variety of sources. Flanagan and Norman (1983) defined these into four subgroups:

- manufacturers
- suppliers and contractors
- modelling techniques
- historical data.

The use of cost data in the cost planning of PFI/PPP projects is directly related to the stage of the cost planning process. For example, in the early stages of the costing process, estimates are based on coarse information, such as functional unit rates. These estimates are progressively developed until the FC stage where detailed data on elemental cost analysis are required for pricing with certainty the capital costs.

The data required for cost planning can be derived from a range of possible sources (Bennett and Ferry 1987):

- direct estimation from known costs and components;
- historical data from typical applications;
- models based on expected performance, average, etc.;
- best guesses of the future trends in technology, market application;
- professional skill and judgement.

Whatever the source of data, a thorough knowledge of the data used in the cost planning process should be acquired to ensure that the techniques used are valid and are not likely to introduce biases into the cost modelling process.

Defining and developing cost assumptions

All aspects of cost planning are based on service requirements, economic, financial and service lives assumptions. Appropriately defined and developed assumptions are probably the most important issues in the cost planning of PFI/ PPP projects. They are important because all the computation of the cost plans is based on them. A cost plan that is based on inaccurate assumptions can lead to inappropriate decisions and loss of value. A comprehensive list of assumptions and objectives for whole life budgeting can be found in the research of Boussabaine and Kirkham (2004).

From the client's point of view at the early OBC stage, space and capital cost assumptions are normally developed by architects and quantity surveyors with detailed experience in particular types of projects. The resultant costs are usually subject to detailed objective scrutiny by external cost consultants and by reference to current PFI/PPP schemes relevant to the project under costing.

Both procurer and providers are required to decide on the following key cost planning objectives/assumptions before embarking on the costing exercise of PFI/PPP projects:

- the annual inflation rate;
- the NPV date;
- the discount rate to be used in calculating NPVs;
- period of cash flow occurrence, e.g., all cash flows to occur at the end of each semi-annual period for the purposes of calculating NPVs;
- setting a date for financial close;
- setting a date for the contract expiry, e.g., 35 years from financial close;
- indexation of the service payment, e.g., the service payment will be indexed annually at the RPI on 1 January each year;
- setting an annual service payment indexation factor;
- setting a level for all financing fees;
- the underlying cost of funding assumptions, e.g., the funders' margin;
- assumptions on accounting and tax treatment;
- setting an acceptable project IRR before financing and tax in both real and nominal terms;
- deciding on return on equity and sub-debt in both real and nominal terms and a blended equity return, that incorporates all sub-senior debt finance;
- determining an appropriate annual debt service cover ratio (ADSCR), loan life cover ratio (LLCR) and project life cover ratio (PLCR) for each period of each loan and in aggregate with minimum and average ratios;
- service lives of components and equipments;
- design and construction costs;
- facilities management services expenditure;
- maintenance, repairs and debt service reserve accounts;
- disposal or end-life cost assumptions.

Contingency planning

In traditional cost planning, the objective of contingency planning is to ensure that the estimated project cost is realistic and sufficient to contain any cost incurred by uncertainties. The most widely used method for allowing for uncertainty is to add a percentage sum to the most likely estimate of the final capital costs. It is stated that there is a tendency for estimators to include an inflated buffer in the contingency estimate (Picken and Mak 2001). Recently, considerable attention has been given to the tendency of those appraising PFI/PPP schemes, and those engaged in bidding for its operation, to overestimate the benefits and to underestimate the costs and the risks associated with delivery. In public sector cost planning jargon this has come to be referred to as optimism bias. Optimism bias is the demonstrated systematic tendency for appraisers to be over-optimistic about key PFI/PPP cost parameters. Some authors attribute bias to personal risk attitudes (Raftery 1994). Therefore, allowing for an appropriate contingency in the financial and economic appraisal is very important in PFI/PPP procurement. That is why public project appraisers are normally required to review all the contributory factors that lead to cost and time overruns, as identified by Boussabaine and Kirkham (2004). Optimism bias is estimated in all PFI/PPP appraisals, and can arise in relation to the following:

- capital costs
- whole life costs
- operational costs
- project financial parameters, e.g., interest rate, inflation
- project duration
- development and construction duration
- service life of components
- affordability
- unitary payments
- overstating or understating the benefits and the benefits shortfall.

To minimise the level of optimism bias in appraisal, we advocate that the following actions should be taken:

- The cost planning process should be managed by competent and experienced managers.
- Development of robust costing assumptions.
- Checking and updating all cost assumptions at different key milestones and sign off gateways to make sure that circumstances have not changed and assumptions are still valid.
- Cost sign-off gateway stages should be set up.
- Development of cost breakdown structures and targets.

The reader can consult Boussabaine and Kirkham (2004) for a generic step-by-step framework for developing whole life cost plans. For in-depth detail and an example of how to deal with optimism bias, see Chapters 8 and 12.

Managing the costing process

In complex PFI/PPP schemes, cost planning requires an effective project structure. That structure has to provide a clear chain of reporting and set out the areas of responsibility and the level of decision-making, and cost plan sign-offs. It is important that clients put in place procedures to monitor whether cost plans and quality are maintained post-FC and during construction.

Cost management of PFI/PPP projects begins with a thorough understanding of what activities and events generate costs; only after identifying activities that generate costs can cost management efforts be successful. Management can only impose control by altering or changing the nature and extent of PFI/PPP project aspects that create costs or alternatively do not add value. As demonstrated in Figure 7.1, cost management of PFI/PPP schemes is an evolutionary rather than a static process. Over time (depending on the stage of the cost planning process), costs change for a variety of reasons.

Looking at cost planning management from a whole life cycle perspective, five different cost management activities can be identified. These are:

1. *Project cost management*: managing the costs of PFI/PPP schemes to ensure that the scope, quality, and budget are aligned and to monitor and manage the balance of these three components throughout the development construction phases
2. *Operational cost management*: managing the costs of service provision and conducting evaluation of facilities management's performance
3. *Investment and evaluation*: justifying and prioritising investment proposals through economic and financial analysis
4. *Budgeting*: determining a budget for every aspect of the proposed development, including affordability
5. *Benchmarking*: setting success factors and comparing operational costs to key indicators

Figure 7.2 provides an outline of the framework for managing the cost planning process in PFI/PPP projects. The stages of the cost plan life cycle are condensed into three main categories. These are detailed in the following subsections.

Pre-financial close stage

At the FC stage it is the procurer's responsibility to ensure that he continues to be satisfied with design evolution as well as financial and legal closure. The pre-financial close includes all the cost planning activities that are described above.

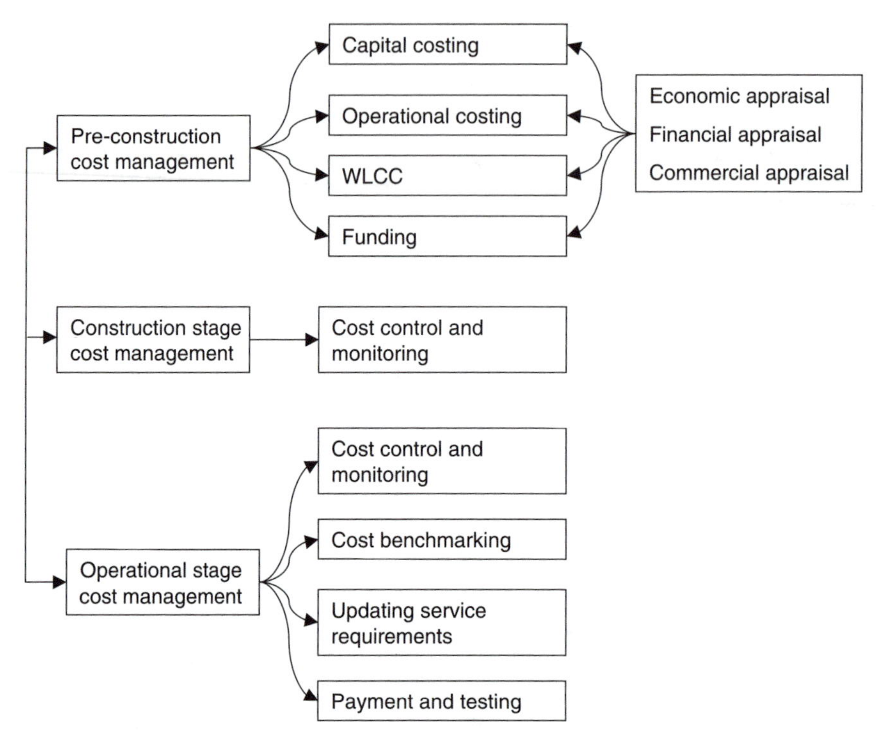

Figure 7.2 A framework for cost planning management.

The major cost management activities in the pre-construction phase involve establishing the cost objectives and targets, based on design and service requirements, putting in place funding mechanisms and carrying out financial and economic analysis (the client's responsibility). The provider's cost management plans include aspects of capital costing including cost estimation, whole life cycle cost forecast and pricing risks. From the FITN stage through to the FC stage, it is the responsibility of the bidders to design, specify price and document the project in accordance with the clients' service requirements. Prior to the INT stage, the majority of cost planning activities and management are carried out by the client and their cost consultants to develop OBC. In complex schemes, it is common practice for both provider and client to employ cost and finance consultants to develop and check the accuracy of their cost models. These consultants also play a vital role in the continuous updating of project estimates and provide feedback on budget impacts of decisions on major service requirements, design aspects and funding issues. This process will continue right up to the FC stage.

In order to manage this process effectively, an important element of cost planning management at this stage is the development of cost breakdown structures and targets. The idea here is to break down the PFI/PPP project's cost into hierarchical cost categories. These are generally established along the line of the cost assumptions and cost activities described above and in the FITN stage. Cost breakdown structure here is essential for management, reporting and control purposes. It is also important for identifying the impact of cost change in a particular area without affecting the other cost categories. Additionally, it is imperative that cost elements that are significant or associated with high risk must be emphasised in the cost planning management control and reporting systems. Value engineering and value management methods ought to be considered at this stage. Any changes to the service requirements, design and funding mechanisms at this early stage may have very little impact on the overall PFI/PPP project overall cost, whereas the benefits from such analysis could be huge for all stakeholders.

Construction stage

During construction, the focus of cost management shifts from predictive pricing to reactive cost control and management of any variation in the work. The construction phase includes all the cost control activities that are required to keep the project to the required budget. The major activities at this stage include monitoring cost targets and construction quality. The provider is responsible for the construction and monitoring progress. The provider must review the construction cost plan against progress achieved (cost to date) in order to understand where cost overrun difficulties are likely to arise and to establish what alternative courses of action are available to mitigate their impact, especially where the overall objectives of the scheme are at risk. It is important here that the providers should assess the effect of construction progress on expenditure cash flow and review any financial implications of any change to the construction programme.

From the client's point of view, during the construction phase the works are normally inspected regularly by the independent tester (IT) and the progress is observed by representatives from the client's organisation to ensure the building meets agreed contract requirements and that the agreed site access arrangements are in place as normally dictated under the project agreement.

To provide data for future cost management, post-construction evaluation is essential. A project evaluation needs to be carried out to prepare a detailed cost analysis of the completed project and to develop lessons learned to inform future PFI/PPP decisions. Chapter 11 provides more information on cost control during the construction phase.

During the commissioning phase, both the provider and procurer are responsible for bringing the new facility smoothly into operation. The commissioning activities are addressed in detail in Chapter 12.

Operation stage

The major cost management tasks in the operation phase include updating service requirements, operation cost monitoring and control and service deliver monitoring and testing. Most of the project costs are incurred during the actual operation stage. It is important that serious planning and control effort is aimed at attaining the service requirements and value for money during the operation phase. Basic cost planning for this phase should be done at the pre-construction stage. This includes establishing whole life costs, service cost and quality, incentive/penalty schemes and developing cost management and reporting systems.

At the operation stage, the provider is required to self-monitor and report PFI/PPP service delivery and performance against the service standards listed in the payment mechanism. The client is normally responsible for sample monitor procurer performance and the service standards achieved. At the operation phase, continuous monitoring and control are necessary to make sure that different service requirements are provided according to the specification and cost targets. To deliver this philosophy, proper monitoring and control procedures must be put in place. This should include cost benchmarking and critical success factors. Benchmarking services that deliver in terms of cost and quality are important for judging the success of PFI/PPP projects in terms of value for money. Benchmarks should be associated with service performance measures. Stakeholders should be able to generate statistics on key performance indicators. Based on these, action should be taken to achieve the planned operational cost and quality objectives.

Occupancy and post-occupancy evaluation is important for future cost management; an evaluation ought to be carried out to prepare a detailed cost analysis of the completed project and to develop lessons learned to inform future PFI/PPP project operation decisions. Chapter 11 discusses operational aspects of PFI/PPP projects. The difference between traditional cost planning and PFI/PPP is that the latter is driven by service output specification requirements rather than capital cost planning.

Summary

The cost planning of PFI/PPP is mainly driven by legally defined documentations like project agreements. The provider's cost planning process is concerned with capital appraisal, financial analysis and affordability studies to develop public sector comparators. The provider's cost plans evolve between the SOC and FBC stages. The cost plans are frozen at the financial close. The provider's cost planning varies from project to project. Providers' cost plans are based on the conditions stated in the bid documentations. In general, they include capital, operation, life cycle and financial cost plans.

Chapter 8

Financial, economic and commercial appraisal at the pre-ITN stage

Introduction

PFI/PPP projects are initiated by the relevant departments of public authorities. For the purpose of checking their economic viability and their adherence to the goals of value for money and affordability agenda and objectives, projects need to be appraised at very early stages by the procuring departments. This chapter sets out the principles for developing the economic case for the proposed PFI/PPP schemes. The process is based on the options appraisal that is required in preparing the OBC and an up-to-date economic analysis to compare the proposed PFI/PPP schemes with a public sector comparator (PSC).

Costing the PFI/PPP project brief

This section introduces the process for costing PFI/PPP project options in terms of both capital and revenue implications. The general approach to the cost appraisal of the schemes from the client's point of view is illustrated in Figure 8.1. As shown in this figure, the service key assumptions are normally used to derive the comparative cost implications of each of the options. Hence, the assumptions are used to obtain and model revenue costs in terms of capital charges, workforce requirements, facilities and any other associated revenue costs. The assumptions are also used to develop capital cost plans. Consequently, the key assumptions and the outcome from the revenue and capital costing are used to carry out financial, economic and benefits appraisal. Subsequently, these costs are adjusted to reflect the risks that will vary over the options. The financial appraisal is the ultimate determinant of affordability while the economic appraisal determines the value for money provided. The commercial appraisal sets out the procurement process, the resulting contract and its key elements, risk transfer and payment mechanisms. The process shown is Figure 8.1 is explained and demonstrated by a case study in this chapter.

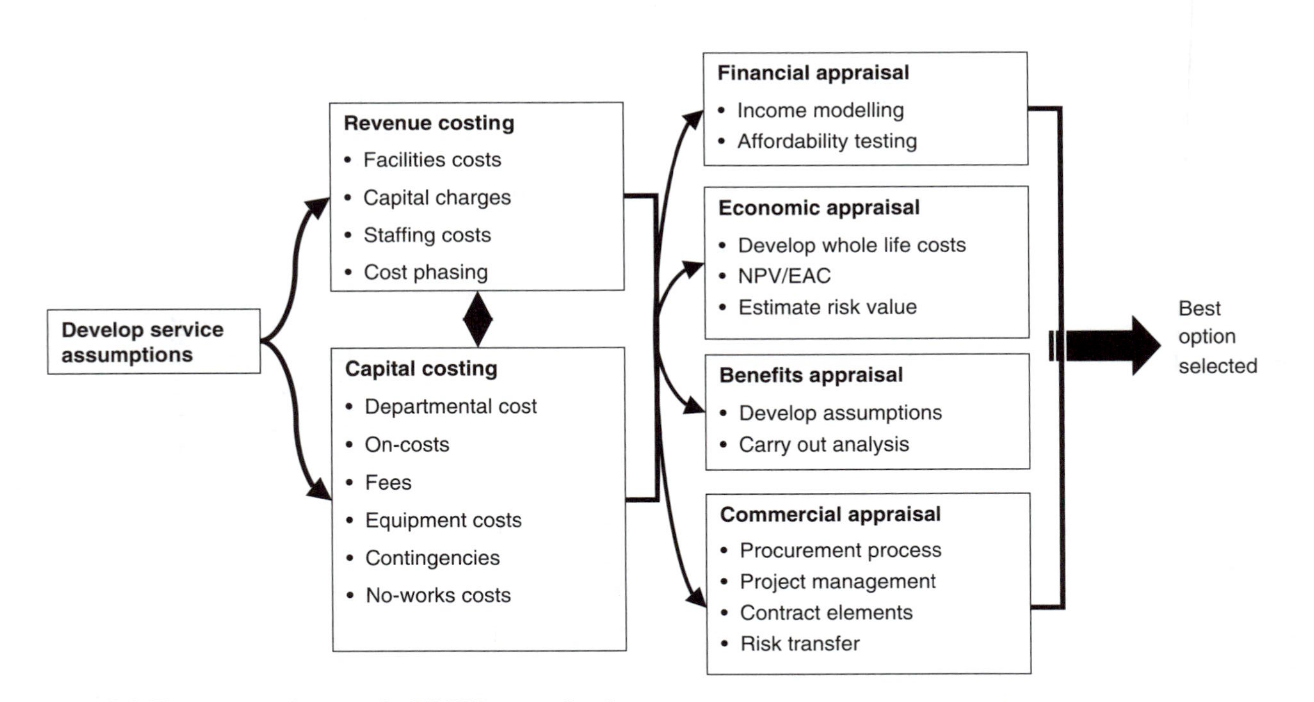

Figure 8.1 The process of costing the PFI/PPP project brief.

Establishing the economic feasibility of PFI/PPP projects

The economic feasibility of investment options is normally driven by key assumptions that have a material impact on the overall operation costs of the scheme under analysis. These assumptions are based on the following aspects:

- the level of capital cost required, which depends on the functional content and services required;
- the level of base costs. This aspect is related to the revenue that is generated from pay and non-pay aspects of the existing services which are to be maintained.
- the level of variations in income, dependent largely on the number of in-patient days, in the case of hospitals;
- the level of variation in the revenue costs;
- forecast capital charges;
- Value Added Tax (VAT) is assumed to be charged at the current standard rate of 17.5 per cent.

The above assumptions, in addition to risk transfer to the private sector, will play a pivotal role in selecting the best investment options. These issues are discussed in detail in the following subsections.

Economic appraisal

The purpose of the economic appraisal is to rank the options in terms of the relative cash impact of the scheme on the procurer's overall financial status, taking into account the timing of cash flows occurrence and the cost of capital.

To demonstrate the concept of economic appraisal, the procurement of a new surgical block is used as an example. The functional content, space requirements, floor area and new build cost per m^2 are shown in Table 8.1 Costing data are extracted from similar schemes.

Schedule of accommodation

In traditional procurement the schedule of accommodation is normally developed from floor plans during the scheme design stages. However, at the early stages of PFI/PPP projects development, the functional content and space requirements are extracted and calculated, based on department activities and service needs. Standard guidance, such as hospital bed per m^2, is usually used for this purpose. In the example shown in Table 8.1 the schedule of accommodation is based on a typical acute non-teaching surgical unit. It is assumed that 50 per cent of beds should be in single rooms. The size of the unit is based on likely demand growth for surgical services in a hypothetical area. In this case

Table 8.1 Departmental and equipment feasibility cost plans for a surgical scheme

Functional content	*Space requirements*	*Area m² (1)*	*New build cost/m² (2)*	*Capital cost £(1) × (2)*	*Equipment Cost £*
General surgery ward 1	50% are single rooms – 40 beds	1380	1500	2070000	144632
General surgery ward 2	50% are single rooms – 30 beds	1050	1500	1575000	108474
General surgery ward 2	50% are single rooms – 30 beds	1050	1500	1575000	108474
Intensive care ward	15 beds	1380	1500	2070000	126766
High dependency	10 beds	293	1600	468800	355245
General theatres	10 theatres	3090	1400	4326000	943290
Surgical investigation units	5 beds	194	1300	252200	159300
Investigation units	5 rooms	1140	1300	19152000	159300
Ultrasound unit	1 room	30	900	27000	189300
Outpatients clinic	10 rooms	600	1000	600000	70000
Consultation rooms	10 rooms	180	1000	180000	50000
Pain clinic	2 rooms	160	1000	160000	150000
Offices – surgical	70 work stations	730	900	657000	286000
Offices – anesthetic	30 work stations	370	900	333000	128800
Offices – admin	1 office	30	950	28500	9000
Offices – secretaries	1 office	25	950	23750	7000
Offices – seminar rooms	50 person	105	1100	115500	10000
Restaurant	150 seats	550	800	440000	100000
Coffee lounge	flexible	200	1200	240000	300000
Central kitchen	1000 meals	1000	1700	1700000	1000000
Support – cleaning	1 store	20	800	16000	0000000
Support – kitchens	4 rooms	212	1400	296800	300000
Support – changing rooms	female 2 areas	130	900	117000	7000
Support – changing rooms	male 2 areas	130	900	117000	7000
Support – visitors WCs	10 toilets	50	1000	50000	2600
Support – IT rooms	5 rooms	70	1000	70000	625000
Support sterile services	5 sterilisers	1500	1700	2550000	300500
Total		15669		39210550	5647681

study, the in-patient days, i.e. the period of admission until discharge is assumed to be around 70,000 days per year. Table 8.1 also displays the floor area of proposed functional content spaces. The total build floor area is 15669 m^2.

Departmental costs

In addition to the schedule of accommodation, Table 8.1 provides the cost/m^2 for new build and capital costs for each of the proposed functional content. The functional content of the hypothetical unit is costed based on the UK NHS standards provided in the NHS Executive's Healthcare Capital Investment cost guidance, together with the areas determined from the schedules of accommodation. The Departmental Cost Allowance Guide (DCAG) provides for only new build projects costs. If refurbishment is required to upgrade existing buildings, then it is safe to assume abatement between 0.25 of new build for minimum refurbishment to 0.75 or more for significant refurbishments. The total departmental costs, including VAT, shown in Table 8.5 are 63 per cent of the project works cost. This is comparable with similar schemes. For control, updating and auditing purposes, it is important to index these costs. In the example illustrated in Table 8.5, departmental costs are indexed at MPIS 4Q05 level.

Works cost

Table 8.2 provides estimated prices for non-standard departmental cost, such as circulation, lifts, etc. These costs are extracted from published rates for similar type of projects and departments that have similar characteristics. Of course, these costs are subject to regional variations, specification of materials and quality of workmanship. External engineering works cost is derived by applying a percentage rate to the departmental cost as shown in Table 8.2. These percentage rates are normally based on schemes of similar size and nature. The rates in Table 8.2 are used to compute the total on-costs shown in Table 8.5, column 2. These costs are also subject to indexation as shown in Table 8.5. The estimated total works cost for this scheme is 36 per cent of the project works cost. The forecast outturn shown in row 12 of Table 8.5 forms the initial capital investment costs required to develop the hypothetical surgical unit.

Regional location factor

It is accepted that building costs are sensitive to regional variations. In the example shown in Table 8.5, the works cost is adjusted upwards by 3 per cent. This is just a theoretical value, but in a real scenario, cost data and location factors can be used for such computation. For example, in the UK such values can be taken from the NHS *Estates Quarterly Briefing* publications or other published location factors such BCIS. In setting location factors it is important that the analyst should consult previous projects in the area to check the validity

Table 8.2 On-costs plan for the proposed unit

Item description	*Quantity*	*Unit*	*Rate*	*Cost £*	*% of on-cost*
1 Communications					
a Space					
Building Costs					
Hospital corridors average rate; part glazed	4000	m^2	1200	4800000	
Other interdepartmental circulation space	700	m^2	800	560000	
Stair lobbies	80	m^2	900	72000	
Stairs (per flight) –	10	Nr	20000	200000	
Lift shafts	55	m^2	500	27500	
Lift lobbies	30	m^2	900	27000	
Plantrooms – Internal – Roofspace	2500	m^2	600	1500000	
Vertical ducts	500	m^2	500	250000	
Total				7436500	33.4
Engineering Costs					
Stairs and corridors/circulation	300	m^2	400	12000	
Plant room/walkway services	200	m^2	500	100000	
Total				112000	0.5
Main contractors discount/profit, prelims and contingency	included				
b Lifts					
New two-bed lift	1			150000	0.68
Main contractors discount/profit, prelims and contingency	included				
2 External Building Works					
a Drainage					
Foul and surface water to buildings	20000	m^2	14	280000	
Surface water to paved areas	item			40000	
Alterations to surface water	18000	m^2	15	324000	
Final connections	item			5000	
CCTV drainage inspection	item			50000	
Total				699000	3.15
b Roads, Paths and Vehicle Parking					
Minor works to existing car parking Item	item				
Realignment of site access road into service area	2000	m^2	50	100000	
Delivery areas	500	m^2	60	30000	
Paths/paving	1000	m^2	40	40000	
Landscaping car park areas	2000	m^2	20	40000	
Total				210000	0.94
c Site Layout, Walls, Fencing, Gates, etc.					
Landscaping to courtyards	1000			50000	
Landscaping to remainder of site	5000	m^2	50	100000	
Fencing/gates	2000	m^2	20	30000	

Site furniture	item	m^2	15	10000	
External signage	item			10000	
External steps	100			10000	
Walls/retaining walls	item	m^2	100	60000	
Total Sundries				270000	1.24
3 External Engineering Works					
a Steam, Condensate, Heating, Hot Water and Gas Mains		%DC	4%	1568422	7.05
Budget allowance based on cost norms					
b Cold Water Mains and Storage		%DC	2%	784211	3.5
Budget allowance based on cost norms					
c Electricity Mains, Sub-Stations and Standby Generating		%DC	5%	1960527	8.8
Budget allowance based on cost norms					
d Calorifiers and Associated plant		%DC	2%	784211	3.5
Budget allowance based on cost norms					
e Miscellaneous		%DC	10%	3921055	17.6
Budget allowance based on cost norms					
4 Auxiliary Buildings					
Bin stores, cycle sheds, etc.		item		15000	
Sub-stations (housing only)				20000	
Generator house (housing only)				25000	
Total				60000	0.26
5 Abnormals					
a Building					
Abnormal foundations/ground conditions	500	m^2	60	30000	
Art works	1000	m^2	20	100000	
Demolition/alteration works at interface of new/existing				130000	0.6
Demolitions	2000	m^2	80	160000	
Decontamination of ground following demolitions		Item		30000	
Asbestos removal in plant spaces		Item		50000	
Access difficulties/temporary works				240000	1.08
b Engineering		%DC	10	3921055	17.6
Budget allowance based on cost norms					
Total on-costs				22246981	

Source: Adopted from National Health Service Estates (1995)

of the level of factor selected. Taking an average of published location factors for the area could be a good compromise and may yield better results.

Fees

An allowance of 23 per cent of the works cost has been made for fees, this is to cover legal, financial, and town planning advice, project management costs, the appointment of a project director, appointment of a commissioning officer, preparation of approval documents, design team fees, quantity surveying, survey and report costs, clerk of works and planning supervisor fees. The breakdown of these fees is presented in Tables 8.3 and 8.4. The level of fees may change according to the complexity of projects.

Non-works costs

The proposed site for development requires further work before the actual construction process starts. Hence, abnormal costs are identified in Tables 8.3 and

Table 8.3 On-costs cost plan summary

Item description	*Estimated cost (exc. VAT) (£)*	*Percentage of departmental cost*
1 Communications		
a. Space	7548500	19.25
b. Lifts	150000	0.38
2 External Building Works		
a. Drainage	699000	1.78
b. Roads, paths, parking	210000	0.54
c. Site layout, walls, fencing, gates	270000	0.69
d. Builders' work for engineering services outside buildings		
3 External Engineering Works		
a. Steam, condensate, heating, hot water and gas supply mains	1568422	4.00
b. Cold water mains and storage	784211	2.00
c. Electricity mains, sub-stations, stand-by generation plan	1960527	5.00
d. Calorifiers and associated plant	784211	2.00
e. Miscellaneous services	3921055	10.00
4 Auxiliary Buildings	60000	0.15
5 Other on-costs and abnormals		
a. Building	370000	0.94
b. Engineering	3921055	10.00
Total on-costs	22246981	56.74

Source: Adopted from National Health Service Estates (1995)

Table 8.4 Non-works and fees cost plan

Item description	*Costs (£)*	*Percentage of works cost*
1 Fees (including 'in-house' resource costs)		
a Architects	3424252	5.5
b Structural engineers	1556478	2.5
c Mechanical engineers/electrical engineers	2179070	3.5
d Quantity surveyors	933887	1.5
f Project management	933887	1.5
g Project sponsorship	311295	0.5
h Legal fees	311295	0.5
i Site supervisor	933887	1.5
j Planning supervisor	311295	0.5
k Reports	311295	0.5
l Expenses/contingency	3424252	5.0
Total fees	14630893	
2 Non-works costs		
a Land purchase costs and associated legal fees		
b Statutory and local authority changes	93389	0.15
c Building regulations and planning fees	186777	0.3
d Decanting costs	186777	0.3
e Temporary accommodation (offices)	249037	0.4
Total non-works costs	715980	

Source: Adopted from National Health Service Estates (1995)

8.4 and these are related to item of works such as demolition, decontamination of site, etc. An allowance of 1 per cent of works cost is made for this purpose. These items are listed in Table 8.2 under Abnormals. These costs are normally based on a quotation from subcontractors or rates from similar building activities.

Equipment costs

In the majority of PFI/PPP health schemes, the equipment cost is derived from NHS standards provided in the NHS Executive *Capital Investment* (1994) cost guidance together with the areas determined from the schedule of accommodation. However, if equipment costs are unavailable in the standards, these might be based upon experience of the development of similar projects. If equipment is purchased by the client from normal revenue budgets, e.g. surgical instruments, desktop equipment, etc., it is imperative that these costs must be excluded from the capital costs. Details of the equipment costs shown in Table 8.1 are extracted from recent NHS buildings and they are also uplifted to reflect the change in the price index as shown in Table 8.5.

Planning contingency

Chapter 7 provides a detailed account on issues that need to be considered in planning contingency. In this case study, a contingency of 15 per cent on the total of works cost, fees, non-works cost and equipment costs is estimated to cover any incidentals, as shown in Table 8.5. However, the procurer should not rely exclusively on these to control cost variations. Procurers should make all possible efforts to effectively manage the project within budget. In reality, contingencies are assessed and costed individually for each option. The capital cost shown in Tables 8.1–8.5 has excluded cost adjustment for optimism bias. These issues are dealt with in Chapters 11 and 7.

Indexing

The economic appraisal of PFI/PPP schemes can span over many months and in most cases years. Hence, is it vital that all estimated prices are subject to inflation adjustment, preferably at each key stage of the appraisal process and most appropriately at each sign-off gateway. Presumably at each of these key milestones most prices included in the economic appraisal are not the current market value, and they should be adjusted accordingly to reflect future inflation. In the example shown in Table 8.5, an indicative inflation adjustment is made in rows, 3, 8 and 11. This is split between inflation up to present day (based on MIPS and equipment index), inflation up to financial close (based on MIPS and equipment index) and equipment inflation up to purchasing date. For example in Table 8.5, row 11, inflation adjustment is made in order to bring construction cost to current date by subtracting the equipment cost from the total at row 10 and then multiplying by the MPIS index, thus:

$$\text{Uplift construction cost to MIPS 471} = \frac{(10\text{b} - 8)(485 - 466)}{466}$$

Cash flow

Cash flows relating to risk are evaluated because under the proposed PFI/PPP project, significant risks would be transferred from the procurer to the provider. Therefore, it is imperative to identify differences to the client's exposure to risk under the PSC and the PFI scheme as part of the VFM analysis. For in-depth detail of PFI/PPP schemes cash flow, readers should refer to Chapter 6.

The cash flows that are included in economic analyses are mainly relating to the payments to acquire the capital assets. In the majority of cases, these cash flows are assumed to fall in the early years of the period under analysis. This period may extend up to the completion of construction and purchase of equipment stage, in some cases, it may extend beyond these stages. Hence, for the purpose of cash flow and whole cycle computation of capital expenditures, PFI/PPP schemes could be assumed to spread over a period of up to ten years

Table 8.5 Capital costs plan summary

	Budget description	*Cost exc. VAT (£)*	*VAT (£)*	*Cost inc. VAT (£)*	*Percentage of works cost*
1	Department costs DC (from Table 8.1)	39210550	6861846	46072396	63
2	On-costs (a) (from Tables 8.2 and 8.3)	22246981	3893222	26140202	36
3	Works cost total (1 + 2) at FP MIPS 4Q05 460 adjusted to 1Q06 466 MIPS (Tender price index level 1995 = 100 base)	61457531 801612 62259143	10755068 140282 10895350	72212598 941894 73154493	
4	Provisional location adjustment (if applicable) (3 % of subtotal 3)	1843726	322652	2166378	3
5	Sub-total (3 + 4)	64102869	11218002	75320871	
6	Fees (from Table 8.4)	14630893	2560406	17191299	23
7	Non-works costs (from Table 8.4)	715980	125297	841276	1
8	Equipment cost (from Table 8.1) 150 ECAG @ 4Q05 160 uplifted to 1Q06	5647681 6024192	988344 1054234	6636025 7078426	10
9	Contingencies on 5, 6, 7 and 8 @ 15%	12821090	2243691	15064781	21
10	TOTAL (5 + 6 + 7 + 8 + 9)	98295024	17201629	115496653	21
10a	Residual optimism bias @ 13% of 10	12778353	17201629	15014564	
10b	Total at MIPS 466 (including 10a)	111073377	19437841	130511218	7
11a	Inflation adjustments Construction to current day – MIPS 485.62 2Q06 Equipment no change required	4422416 000	773923	5196339	
11b	Total at current prices 1Q06 MIPS 466	115495793	20211764	135707556	
11c	Inflations adjustments: Construction to FC MIPS 483 1Q07 During construction NA Equipment to purchasing date 170 1Q09	4469867 376512	782227 65890	5252094 442402	7 1
12	FORECAST OUTTURN TAKEOVER BUSINESS CASE TOTAL	120342172	21059880	141402052	

Source: Adopted from National Health Service Estates (1995)

depending on the complexity and size of project under analysis. The cash flow profile of the scheme should also include all annual cash flows related to the lease rental payments. In normal circumstances, these should be assumed to occur every year up to the end of the concession period. Capital receipts, opportunity costs and residual values should also form part of the cash flow analysis. Running costs, i.e., capital charge, service delivery costs and premises costs, are included in the analysis. Based on these cost flow assumptions Table 8.7 is derived to compute the NPV of the estimated cash flows for this case study. The expected cash flow of the capital costs is shown in Table 8.8.

Whole life cycle costs

The economic appraisal requires the whole life cycle costs for PFI/PPP schemes. At the early stages of project development, building components and services specification have not yet been specified. This will make it almost impossible to specify with certainty the building components service lives which are essential for development of whole life cycle maintenance and replacement budgets. The whole life cycle estimates shown in Table 8.6 are based on previous experience of other health-care schemes of similar size, function and characteristics. The estimates included in Table 8.6 are based on a percentage of works costs. The estimates include an allowance for reactive and planned maintenance and accompaniments replacement. Notice in Table 8.6 that the allowance percentages increase at certain key dates to allow for extensive planned maintenance. Notice also towards the end of the concession that the maintenance and repayment allowances are significantly high. It is important for whole cycle evaluators to recognize that the allowance for whole life cycle costs varies according to the age and type of the building. If refurbished buildings are appraised as part of an option, then these allowances should be substantially increased. Hence, a sliding increased rate is used for this case. It is also perfectly reasonable to use different assumptions for driving whole life cycle costs. For example, analysts may estimate these costs as equivalent to the initial capital cost excluding equipment and contingency sums but including fees, spread in equal instalments from the start to the end of the concession agreement. For further details, refer to Boussabaine and Kirkham (2004).

Net present cost calculation

Net present cost is determined by discounting all the cash flows associated with each PFI/PPP scheme option back to their present value. The discounted cash flow method is based on the paradigm that more weight is given to earlier costs than to later costs. This is taking into account the different timings of cash flows. From a decision-making point of view, this might be interpreted as it were preferable to pay costs later rather than sooner, and to receive benefits sooner rather than later. A discounted cash flow for the hypothetical case is

Table 8.6 Project maintenance and replacement costs over the appraisal period

Year	*DF @3.5%*	*Maintenance*			*Replacement*		
		% works cost	*forecast*	*NPV*	*% works cost*	*forecast*	*NPV*
1	0.9662	0.2	124518	120310			
2	0.9335	0.2	124518	116238			
3	0.9019	0.2	124518	112303			
4	0.8714	0.2	124518	108505			
5	0.842	0.55	342425	288322			
6	0.8135	0.21	130744	106360			
7	0.786	0.2	124518	97871			
8	0.7594	0.2	124518	94559			
9	0.7337	0.3	186777	137039			
10	0.7089	0.6	373555	264813	0.9	560332	397220
11	0.6849	0.2	124518	85283	0.9	560332	383772
12	0.6618	0.2	124518	82406			
13	0.6394	0.5	311296	199042			
14	0.6178	0.2	124518	76927			
15	0.5969	0.2	124518	74325	15	9338871	5574372
16	0.5767	4	2490366	1436194	8	4980731	2872388
17	0.5572	0.6	373555	208145			
18	0.5384	0.2	124518	67041			
19	0.5202	0.3	186777	97162			
20	0.5026	0.3	186777	93874			
21	0.4856	3	1867774	906991			
22	0.4692	0.3	186777	87636			
23	0.4533	0.3	186777	84666			
24	0.438	0.3	186777	81809	0.5	311296	136348
25	0.4231	0.6	373555	158051	0.5	311296	131709
26	0.4088	0.2	124518	50903			
27	0.395	0.2	124518	49185			
28	0.3817	0.3	186777	71293			
29	0.3687	0.3	186777	68865	5	3112957	1147747
30	0.3563	6	3735549	1330976	28	17432560	6211221
		Totals	13111776	6757094		36608376	16854777

undertaken over 30 years; results are shown in Table 8.7, using a discount rate of 3.5 per cent.

The NPV results shown in Table 8.7 are only valid if the competing PFI/PPP options are of the same life span. However, this is not the case in PFI/PPP procurement, because the PSC does not have the same life span. To overcome this problem, the analyst needs to express the competing options in equal terms. One also should assume that unequal life PFI/PPP options being considered will have the same level of risk. If this is the case, it will be appropriate to use the equivalent annual cost method (EAC) also known as 'Annual Equivalent Annuity Method' to compare net present values of costs on an annual basis.

EAC computes the present value of costs for each PFI/PPP option over a cycle and then expresses the present value in an annual equivalent cost using the

Table 8.7 Discounted cash flow appraisal

Year Nr	*Year*	*to Year*	*Discount factors @ 3.5%*	*Capital expenditure*	*running costs*	*total cash outflow*	*NPV of column 4*	*whole life costs NPV*	*Total NPV*
				£'000	*£'000*	*£'000*	*£'000*	*£'000*	*£'000*
			1	2	3	4	5	6	7 = 5 + 6
0	2003	2004	1	4276		4276	4276		4276
0	2004	2005	1	12827		12827	12827		12827
0	2005	2006	1	38483		38483	38483		38483
0	2006	2007	1	29931		29931	29931		29931
1	2007	2008	0.9662		92943	92943	89802	125	89926
2	2008	2009	0.9335		92943	92943	86762	125	86887
3	2009	2010	0.9019		92943	92943	83825	125	83950
4	2010	2011	0.8714		92943	92943	80991	125	81115
5	2011	2012	0.842		92943	92943	78258	342	78600
6	2012	2013	0.8135		92943	92943	75609	131	75740
7	2013	2014	0.786		92943	92943	73053	125	73178
8	2014	2014	0.7594		92943	92943	70581	125	70705
9	2014	2015	0.7337		92943	92943	68192	187	68379
10	2015	2016	0.7089		92943	92943	65887	934	66821
11	2016	2017	0.6849		92943	92943	63657	685	64342
12	2017	2018	0.6618		92943	92943	61510	125	61634
13	2018	2019	0.6394		92943	92943	59428	311	59739
14	2019	2020	0.6178		92943	92943	57420	125	57545
15	2020	2021	0.5969		92943	92943	55478	9463	64941
16	2021	2022	0.5767		92943	92943	53600	7471	61071
17	2022	2023	0.5572		92943	92943	51788	374	52161
18	2023	2024	0.5384		92943	92943	50041	125	50165
19	2024	2025	0.5202		92943	92943	48349	187	48536
20	2025	2026	0.5026		92943	92943	46713	187	46900
21	2026	2027	0.4856		92943	92943	45133	1868	47001
22	2027	2028	0.4692		92943	92943	43609	187	43796
23	2028	2029	0.4533		92943	92943	42131	187	42318
24	2029	2030	0.438		92943	92943	40709	498	41207
25	2030	2031	0.4231		92943	92943	39324	685	40009
26	2031	2032	0.4088		92943	92943	37995	125	38120
27	2032	2033	0.395		92943	92943	36712	125	36837
28	2033	2034	0.3817		92943	92943	35476	187	35663
29	2034	2035	0.3687		92943	92943	34268	3300	37568
30	2035	2036	0.3563		92943	92943	33116	21168	54284
Total				85516	2788290	2873806			1844653

Note: Capital expenditure includes works cost, major equipment costs and fees

appropriate annuity factors for each cycle. The annuity factor is given by the following expression:

$$AF = \frac{1}{i}\left[1 - \frac{1}{(1+i)^t}\right]$$

$$EAC = \frac{NPV}{AF}$$

Where:

AF = the annuity factor

i = discount rate

t = the period of analysis.

For example, the AF for the case study presented in Table 8.6 is computed as follows:

$$AF = \frac{1}{0.035}\left[1 - \frac{1}{(1 + 0.035)^{30}}\right] = 18.39205$$

$$EAC = \frac{1844653 \times 10^3}{18.39205} = 100296.3 \times 10^3$$

In the above computation the discount rate is assumed to be at 3.5 per cent and the life span of the project is 30 years. Having calculated the EAC for each investment option, the analyst then compares the EACs. The annual equivalent of NPVs of the PFI/PPP competing options can then be compared. Normally in PFI/PPP evaluation, the option that has the lowest EAC over the concession period or the life cycle of the project (60 years in most cases) is the better one for VFM.

All the above results of the economic appraisal must be subject to a sensitivity analysis to examine the relationship between the overall prices of the scheme and changes to the costing parameters. The process is carried out by changing parameters such as capital costs, operating costs and whole life cycle costs. For this purpose, it is expected that the economic analyst will use a spread sheet to increase or decrease the model's input value per certain percentage to test its effect on the overall cost. The outcome of this process is to evaluate whether the sensitivity would have a material impact on the ranking of the alternative options under consideration.

Financial appraisal

This section provides an assessment of the funding solution, in terms of affordability and impact on the procurer's finances. The impact of PFI/PPP schemes on the client's future revenue income and expenditure position is also discussed. The methodology for costing the capital and revenue implications of the case study is explained. The methodology followed in developing financial appraisal includes three main elements: (1) developing assumptions for revenue costs; (2) income projection; and (3) sensitivity testing.

Identifying revenue costs

Revenue costs are related to work and services performed. The revenue costs associated with most public PFI/PPP schemes fall into four broad categories. These are:

- capital charges;
- service delivery costs, mainly staffing costs;
- supporting services costs, such as portering, catering, cleaning and estates;
- bespoke additional revenues.

In computing the capital charges, the split of capital expenditure into buildings and engineering is assumed to be in line with similar projects. Equipment, in this example, is assumed to have a service life of ten years. The cost breakdown of calculating the capital charges is shown in Table 8.8. Depreciation is derived by dividing the total in column 7 by the service life in column 8. For example, buildings depreciation and interest and capital charges are calculated as follows:

$$\text{Buildings depreciation} = \frac{86120}{60} = £1435$$

$$\text{Interest on building capital costs} = \left[86120 + \frac{(86120 - 1435)}{2}\right] \times 0.035 = £2989$$

$$\text{Capital charges on building costs} = 1435 + 2989 = £4424$$

The revenue costs assumed for the purpose of this case study are set out in Table 8.9. It is important to stress that all revenue costs and income must be based on specific date prices. This will enable the decision-makers to update and cross-reference all relevant costs. Service delivery cost and supporting costs for this case study are derived from NHS estates database from similar hospitals units, see Boussabaine and Kirkham (2006). In Table 8.9 direct clinical costs are derived from the staffing and clinical requirements such as doctors, nurses, specialised teams, drugs, blood services, medical secretaries, diagnostic services, and so on. Site support services are related to catering, site costs, postage, patient transport and overheads.

Identifying income

In order to test any PFI/PPP scheme for affordability, a cost envelope must be based on tariff prices that are appropriate for the sector concerned. Under the agenda of modernisation, public authorities can only fund the revenues costs of capital investment by generating income from increased activity (additional workload with the same resources) or efficiency savings (productivity). For

Table 8.8 Capital charges computation

	Building cost split	*Works costs*	*Fees*	*Non-works*	*Buildings*	*Bias and contingency*	*Total*	*Service life*	*Depreciation*	*Interest 3.50%*	*Total capital charges*
	1	*2*	*3*	*4*	*5 = 2 + 3 + 4*	*6*	*7 = 5 + 6*	*8*	*9 = 7/8*	*10*	*11 = 9 + 10*
		£'000	*£'000*	*£'000*	*£'000*	*£'000*	*£'000*		*£'000*	*£'000*	*£'000*
land											
buildings	65%	55750	11174	547	67471	18649	86120	60	1435	2989	4424
engineering	35%	30019	6016	294	36329	10227	46556	15	3104	1575	4679
equipment					7521	1203	603	10	60	20	80
Total		85769	17190	841	111321	30079	133279		4599	4584	9184
		85769304	17191299	841276	7520828	30079345	141402052				
											9183679

Table 8.9 Revenue costs of the proposed scheme

Cost group	*£'000*
direct clinical costs	65000
paramedical services	900
diagnostic services	9500
site support services	15000
Direct Revenue Costs	90400
capital charge (from Table 8.8)	918
2006/2007 inflation @ 5%	1625
Total revenue costs	92943

example, income in acute health schemes, in the UK, is based on projected clinical activities. The activity is measured in patient spells, which relates to a patient's stay in hospital, and may cover many procedures such as emergency admissions, long-stay admissions, etc. Hence, income in these types of schemes is generated depending on results. The projected incomes assumed for the purpose of this case study are set out in Table 8.10. Consequently, income is derived based on an agreed activity multiplied by a nationally set tariff price per patient spell. This result is sensitive to regional variations. The data shown in Table 8.10 are extracted from similar operating schemes. The data on patient spells are published annually by the UK Department of Health. For the purpose of calculating income, in schemes such as schools and prisons, the number of pupils and inmates may be used. Tariff income for the case study is based on several surgical service specialties. The cost per spell is based on assumptions from similar operating surgical units.

Table 8.10 Projected income based on inpatient/days

Source of income	*Projected spells* *1*	*£/spell* 2	*Income £000* *1 × 2*
Outpatient income			190
General surgery	10000	1800	18000
Urology	6000	900	5400
Gynaecology	4000	1100	4400
Ophthalmology	6000	850	5100
ENT	5000	1100	5500
Orthopaedics	8000	2500	20000
Oral surgery	2000	800	1600
Private patients	1000	3000	3000
Other specialties	25000	1200	30000
Total	67000		93190

Testing for affordability

The output from Tables 8.9 and 8.10 is used to assess the affordability of the new proposed scheme. As shown in Table 8.11, the scheme demonstrates revenue affordability of costs against income. Unfortunately the scheme shows a deficit when all whole life costs and risks are considered against income. However, if the set of income and revenue assumptions changes, this could lead to different conclusions. This is why it is important to subject these assumptions to a vigorous sensitivity testing to demonstrate the variability of revenue and income and to test the robustness of the affordability decisions. Hence, the key variables of the revenue costs must be changed to determine how robust the affordability conclusions are. The variables that should be sensitised are project-dependent. For example, in the present case study, the following variables have been selected for testing, based on their percentage contribution to the total revenue and income:

- *Capital costs*: Normally capital costs would have an impact on revenue costs through capital charges. However, in this case study, the capital charge is only 1 per cent of the total revenue cost of the project and thus has only a minor effect on the total revenue costs. However, there is always a risk that the buildings could be valued upwards at the next sign-off gateway, with a knock-on effect on affordability testing.
- *Site support services*: This group accounts for 16 per cent of the total revenue cost. Any increase in the level of site support in terms of catering, postage, etc., will automatically result in a considerable increase in revenue costs.
- *Clinical costs*: Direct clinical costs account for 70 per cent of the revenue costs. The size of these could vary adversely if more staff are required to deliver the anticipated services.
- *Inpatient/day changes*: As can be seen in Table 8.10, the income of the proposed project is generated from patient activity, if this drops significantly,

Table 8.11 Affordability analysis

Cost group	*£'000*
Based on WLCC, risk and income	
EAC	109466
Risk	1094
Total	110560
Total income	93190
Surplus/deficit	–17370
Based on income and renveue	
Total revenue costs	92943
Total income	93190
Surplus/deficit	247

then the viability and affordability of the proposed scheme will be questionable. There would also be a drop in costs associated with a drop in activity. If we assume a 5 per cent drop in inpatient days across all activities in Table 8.10, this will lead to a huge net income loss.

Commercial appraisal

The commercial appraisal deals with key elements, such as the procurement process, the resulting contract and its key elements, risk transfer and payment mechanisms. Chapters 5 and 6 provide further details on these issues.

It is not a requirement to provide a commercial appraisal at an early stage of the PFI/PPP schemes' development. However, we advocate that procurers should carry out some sort of analysis to inform the evaluation of alternative solutions in the decision-making process. It is important that the client's financial advisers should review the project funding options and the level of the unitary payment with a view to confirming that the proposed funding package represents value for money for the project under consideration.

In addition to the design and construction of the PFI/PPP facilities, over the concession period of the contract, the client needs to clearly identify non-clinical facilities management services to be supplied by the provider. It is also important to define and agree standards against which services are going to be provided.

Evaluating alternative solutions

It is normal practice for the procuring authority usually to identify a list of possible options, for delivering the required services, based on clearly defined sets of objectives and constraints. From a preliminary set of potential alternatives, the procurer in most cases short-lists two or three options, prior to undertaking formal option appraisal. The status quo must also be presented as an option for appraisal. However, in most cases the status quo option has already been identified as unsustainable and it is included in the evaluation process as a benchmark. This process is referred to as public sector comparator (PSC). All short-listed options are then costed in terms of both capital and revenue implications, as demonstrated above. The aim is to derive the comparative cost implications of each of the listed options. Before such a comparison is carried out, all costs must be adjusted to reflect the risks that will vary over the options.

The PSC provides the benchmark against which the value for money of the PFI/PPP schemes is tested. Effectively, the PSC represents the physical solution that the client would adopt if a decision is made to fund a particular scheme using conventional public funding arrangements. To make a fair comparison, the analyst should assume that public funding would be available over the same timescale as for the PFI/PPP scheme. This assumption is only theoretical; in most cases, public funds are not available to finance a PSC. The non-financial benefits of each option should also be considered here.

Having costed all the options, the analyst must then select the one that demonstrates best value for money. In most cases, the initial capital cost estimates are the most important driver in choosing between alternative options. However, it is not always the case that the option offering the best value for money will be affordable by the client, for this reason, affordability should be an integral part of the evaluation process. The affordability issue is a major problem for most purchasers. In some cases, according to Shaoul (2005): clients use the following tactics to bridge the affordability gap:

- transferring land and assets to the SPV;
- increasing or inflating the revenue stream;
- reducing staff expenditure.

Quantifying risks and benefits

The development of capital costs of PFI/PPP schemes, from a client point of view, is demonstrated in the above sections. The financial cost of the scheme must also include the costs of some of the risks associated with design, construction and the operations aspects of the services. An in-depth analysis of risk management in PFI/PPP schemes is presented in Chapters 12 and 13. In most PFI/PPP cases, risk quantification and management are required only in relation to the preferred option (in some cases in all short-listed options). This section is mainly concerned with the adjustment of options cost to reflect the magnitude of risk transfer to the private sector. The economic analysis takes into account adjustments for risk transferred by the project agreement, optimism bias and differential tax take. The risk appraisal exercise involves risk identification, estimation of the probability of the risks occurring and estimation (pricing) of the likely impact of the identified risks. The impact and probability of the risk that may occur can be assessed qualitatively or quantitatively.

For the purpose of the case study presented in this chapter, the estimated probability of the identified risks occurring along with the estimated impact of each identified risk is shown in Table 8.12. In Table 8.12 the average yearly risk and risk value or impact are computed as follows:

$$\text{yearly risk} = \frac{\text{min} + \text{most likely} + \text{max}}{3}$$

$$\text{yearly risk value (impact)} = (\text{yearly risk})(\text{probability})$$

EAC risk value is computed as:

$$\text{EAC risk value} = \frac{20129418}{18.39205} = £1094466$$

Table 8.12 Likely risks and their values

	Risk heading	*Occurrence year*	*Probability*	*Impact*			*Average*	*Yearly risk value*	*NPV of (7)*
				Min	*Most likely*	*Max*			
		1	*2*	*3*	*4*	*5*	*6 = (3 + 4 + 5)/3*	*7 = 6×2*	*at 3.5%*
1	Design risks								
1.1	change in design	1	15	10000	300000	500000	270000	40500	39131
1.2	change in requirements	1	30	10000	300000	500000	270000	81000	78262
1.3	design team default	1	15	10000	300000	500000	270000	40500	39131
2	Construction and development risks								
2.1	cost over-runs	2 to 5	20	20000	500000	1000000	506667	101333	359612
2.2	contractor default	2 to 5	10	500000	1000000	1.5E+07	5500000	550000	1951840
2.3	incorrect cost estimates	2 to 5	25	10000	500000	1000000	503333	125833	446557
3	Availability and performance risks								
3.1	insufficient space and capacity	6 to 30	10	20000	300000	1000000	440000	44000	610592
3.2	unavailability of facilities	6 to 30	5	20000	500000	1000000	506667	25333	351553
3.3	inadequate user environment	6 to 30	5	20000	300000	1000000	440000	22000	305296
4	Variability of revenue risks								
4.1	unexpected changes in allocation of resources	6 to 30	5	10000	500000	1000000	503333	25167	349240
4.2	estimate income is incorrect	6 to 30	25	10000	500000	1000000	503333	125833	1746202
4.3	unexpected changes in service provision policy	6 to 30	20	10000	500000	1000000	503333	100667	1396961
5	Operating cost risks								
5.1	cost overruns on utilities	6 to 30	30	20000	300000	500000	273333	82000	1137922
5.2	inaccurate cost estimates of maintenance	6 to 30	25	20000	300000	500000	273333	68333	8812301
5.3	inaccurate cost estimates of providing services	6 to 30	20	20000	300000	500000	273333	54667	758615
6	Termination risks								
6.1	termination due to default by the procuring entity	6 to 30	5	10000	500000	1000000	503333	25167	349240
6.2	default by the operator (step-in by financiers)	6 to 30	5	10000	500000	1000000	503333	25167	349240
6.3	technical change/obsolescence of building	6 to 30	15	10000	500000	1000000	503333	75500	1047721
				NPV of the total likely risk value					20129418

Generally, risk identification is project-dependent, however, in the majority of PFI/PPP projects, the following key risks are utilised to classify a number of project specific risks:

- planning risks
- design risks
- construction and development risks
- availability and performance risks
- variability of revenue risks
- operating cost risks
- termination risks
- technology and obsolescence risks
- control risks
- residual value risks
- land sales and receipts risk
- equipment risks.

For in-depth analysis of risk quantification, readers can consult Boussabaine and Kirkham 2004). For the purpose of the example shown in Table 8.12, to quantify risk, three possible scenarios are assumed for each risk: minimum impact, most likely impact and maximum impact. For each scenario, an impact cost and its probability of occurrence are estimated. The probabilities of occurrence of risk are multiplied by the impact cost to give an overall value of risk. The expected cost of each risk must also be projected over the periods it would occur and discounted at an appropriate discount rate to give the NPV or NPC (net present cost) for each option. These values are then added to produce an expected risk cost for each short-listed option. The EAC of the risk NPV must be computed and then be added to the overall cost of the scheme before testing for affordability as shown in Table 8.11. It will be good practice if a sensitivity analysis is carried out to check and identify the most influential risk parameters. The data in Table 8.12 can be obtained from brainstorming sessions with project stakeholders, technical workshops, economic and legal advisers and from any hard data or evidence on particular risk impact and their probability of occurrence.

The majority of clients also carry out a benefits analysis to determine whether the proposed PFI/PPP scheme demonstrates value for money. Chapters 2 and 3 have shown that both tangible and intangible benefits should be part of the whole life cycle value creation and exchange. There is no standard process for benefits appraisal; however, the following steps could be used iteratively to derive and quantify the benefits associated with each option.

- identification of the benefits criteria;
- developing weighting and scores of the benefits criteria;

- quantifying the benefits criteria by multiplying the score of each criterion by its weight;
- scoring of the options against the benefits criteria.

The benefits criteria can be obtained from the objectives of the proposed scheme. It is important that these criteria should be scrutinised by all relevant stakeholders, including end users. Such benefits criteria might include issues such as effective provision of services, improving existing services, compliance with safety issues, etc.

The purpose of developing weighting scores is to compare and identify the relative importance of the different benefits criteria impact against each other. There are several ways of developing weightings for the benefits criteria. These could be either quantitative or qualitative. Quantitative methods include expert panel method and monetisation method. The expert method is based on questionnaires, interviews or group discussion of panellists, experts, stakeholders or lay people. The monetisation method is similar to the expert method with the exception that participants are asked to assign monetary value to each benefit criteria. There are also complex weighting methods such as equal weights, rank order centroid weights, rank-sum weights and weighted pairs methods. For the purpose of benefits appraisal most cases use either monetary value or weighted pairs' methods.

Measures for selecting the best PFI/PPP deal

The benefit, economic and financial analyses are normally used to confirm which option provides the best value, affordable project option for the scheme under analysis. The financial appraisal is used as a measure for assessing affordability while the economic appraisal establishes the value for money provided by a particular scheme. Hence, PFI/PPP schemes appraisal depends upon two main measures. First, value for money must be demonstrated and quantified with the use of discounted cash flow techniques. Second, at the same time, the scheme must be affordable to the purchasers. In comparison with VFM decisions, affordability decisions are mostly based on qualitative and operational criteria. In order to select the best PFI deal, VFM is based on the concept of net present cost (NPC), and the sum of the present value (PV) of all costs over the period of interest, including residual values as negative costs (HM Treasury 2004a). If a number of options are being considered, then the option with the lowest NPC will be the most favourable option from a financial point of view. To select the best deal, the whole life cycle costs of the PSC option are discounted to yield an NPC, and compared against the NPC of the preferred option as procured under PFI/PPP.

In most cases the VFM analysis comprises the following elements:

- NPC/NPV of the estimated cash flows for each of the options;
- NPC/NPV of the risk transferred by the PFI project agreement;
- NPC/NPV of the optimism bias related to potential scope-creep of the PSC;
- NPC/NPV of the differential tax-take related to the PFI, which must be added to the PSC.

Based on the results from this calculation, the option with the lowest NPC is selected as demonstrating the greatest financial benefit. Until now it has appeared that the selection is based on a straightforward process. However, this is not the case in real practice. Theoretically, the outcome or the level of NPC is influenced by several factors (Shaoul 2005):

- *The choice of the discount rate*: the choice of discount rate is critical since NPC/NPV comparisons are very sensitive to fluctuation in the discount rate. To address this problem, it is advisable for the analyst to use a series of discount rates to test the sensitivity of the preferred option to different rates. There is a general consensus among economists that the public sector should use lower discount rates than the private sector. The type of discount rate to be used in assessing public investments is also important (see Chapter 5).
- *The context and choice of the PSC*: the robustness of PSC is crucial to any meaningful evaluation of the options. It has to be said that the majority of the PSC cases are badly conceived and developed compared with PFI/PPP options. This could be due to the fact that projects will only receive approval if the PFI/PPP options demonstrate VFM. There is also a huge inconsistency between PSC and PFI/PPP cost assumptions. It is assumed or implied in the computation that the PSC costs should be met in the early years of investment. This is not the case in PFI/PPP option appraisal. From the outset, this type of assumption will put the PSC at a disadvantage in comparison to other investment options. This is due to the fact that the discounting methods favour investment options that defer expenditure over those that have high costs in early years.
- *The risk transferred by the project agreement*: Chapters 12 and 13 deal with the issue of risk in PFI/PPP projects. The whole idea of the PFI/PPP paradigm is based on transferring most of the development and operation risks to the private sector. This will have a colossal impact on the costing of PSC. The more risks that are transferred, the more expensive the PSC becomes relative to other procurement options. Obviously this will lead to the distortion of the comparison. That it is why an optimum risk transfer strategy should be put into place so that the PSC is not disadvantaged against other options.
- *The suitability of discounted cash flow techniques*: many authors have argued against the use of opportunity costs approaches in the public sector.

The argument put forward is that there is a huge difference between public and private sector capital expenditure. The public sector investment decisions are geared towards providing social welfare, whereas the private sector decisions are directed mainly to maximising shareholder value. Others have also questioned the use of NPV rules, such as capital rationing, period of investment opportunities, and the relationship between investment options, in evaluating public sector investments (Edwards and Shaoul 2003).

Summary

This chapter has described and demonstrated financial, economic and commercial appraisal of PFI/PPP projects at the pre-ITN stage from the client's point of view. The process is radically different from traditional project investment appraisal. The financial appraisal is the ultimate determinant of affordability, while the economic appraisal determines the value for money provided. The commercial appraisal sets out the procurement process, the resulting contract and its key elements, risk transfer and payment mechanisms.

Part IV

Cost variations in PFI/PPP projects

Chapter 9

Cost planning at the pre-financial close stage

Introduction

Getting a realistic price is vital to any project success. It is crucial that capital and whole life cycle costs are accurately assessed and are subject to regular reviews at set intervals. In the event that the outturn costs prove likely to exceed the capital cost plan, stakeholders should have at their disposal pre-arranged strategies to deal with and rectify the situation.

Bidders are expected to provide a range of estimates to support their bid submission and assist the purchaser in the evaluation process. Normally, the financial submission requirements are set out in full in the FTIN document and fall into two key parts:

- the financial structure and the deliverability of the funding package;
- the fully costed bid proposals.

The first requirements deal with financial structure and the deliverability of the funding package. This obligation deals with details of the proposed funding structure, setting out clearly the institutions involved, the type of facility to be provided, the amount of funds available and the terms on which these funds are being offered. The majority of these issues have been dealt with in Chapters 4 and 5.

The second requirements, the core of this chapter, deal with the financial projections for the full duration of the concession. This chapter will introduce the cost elements that should be included in a typical PFI/PPP capital cost model. Separate elemental cost breakdowns are provided for each part of the construction works. In addition to this, a detailed and fully quantified cost plan is provided to illustrate the capital cost breakdown. The cost plan is presented in sufficient detail to identify the specification, quantity and rate for all major materials and components adopted for both the builder's work and the engineering installations. The whole life cycle costs are developed and broken down into sub-elements/components. Also a summary cash flow of projected life cycle expenditure (on a year by year basis) is presented.

The PFI/PPP cost planning process

This section is primarily concerned with the process of cost planning of FPI/PPP schemes prior to the financial close stage from the providers' point of view. Setting cost targets at the early stage of development will assist in cost distribution between various parts of the PFI/PPP deal. Traditionally, the cost planning process during the pre-construction stage is interactive and is developed in conjunction with all the design stakeholders. The major key stages of setting capital cost plans are:

- *Briefing*: the aim here is to establish a whole life cycle budget framework for the scheme. The information required for setting targets for the WLC budget is similar to the one presented in Chapter 8 and includes issues such as strategic objectives and context of the investment, functional and operational requirements, and schedule of spaces. WLC budget estimates here are based on rough cost information such as rates per functional unit.
- *Outline proposals*: generally, a WLC budget figure is set at this time based on a number of alternative physical forms of the scheme. It is expected that at this stage a realistic cost will be included for WLCC components such as equipment, running costs, maintenance and replacement, fees, contingencies, etc.
- *Schematic design*: it is expected by this stage that the majority of key physical attributes of the project will have been specified. The WLC budget here must be tested against the initial budgets. VFM and WLCC analyses are carried out to understand the impact of the design decisions on capital costs and on the whole life cycle operational budget.
- *Detailed design*: it is at this phase that design is detailed enough to undertake a full elemental cost analysis to confirm that the project is still within the WLC budget developed in the earlier stages. The cost analysis performed here should lead to the development of a pre-tender estimate for evaluating tenders.
- *Tender stage*: probably this the last sign-off gateway before the construction period. It is imperative that tenders are evaluated and reconciled with the WLC budget. Any errors or misinterpretations ought to be rectified and the WLC budget updated accordingly.

The above traditional cost planning process is linear and excludes a large proportion of WLC budget planning in PFI/PPP schemes. Under PFI/PPP the client pays a private sector partner to raise funds on the client's behalf, to design, run and maintain the scheme. This approach differs radically from the traditional way of cost planning, financing, building and operating public facilities. Figure 9.1 illustrates a typical PFI/PPP procurement, design and cost planning processes. As can be seen from Figure 9.1, providers are involved in the procurement process after the publication of the OJEU notice. Their cost

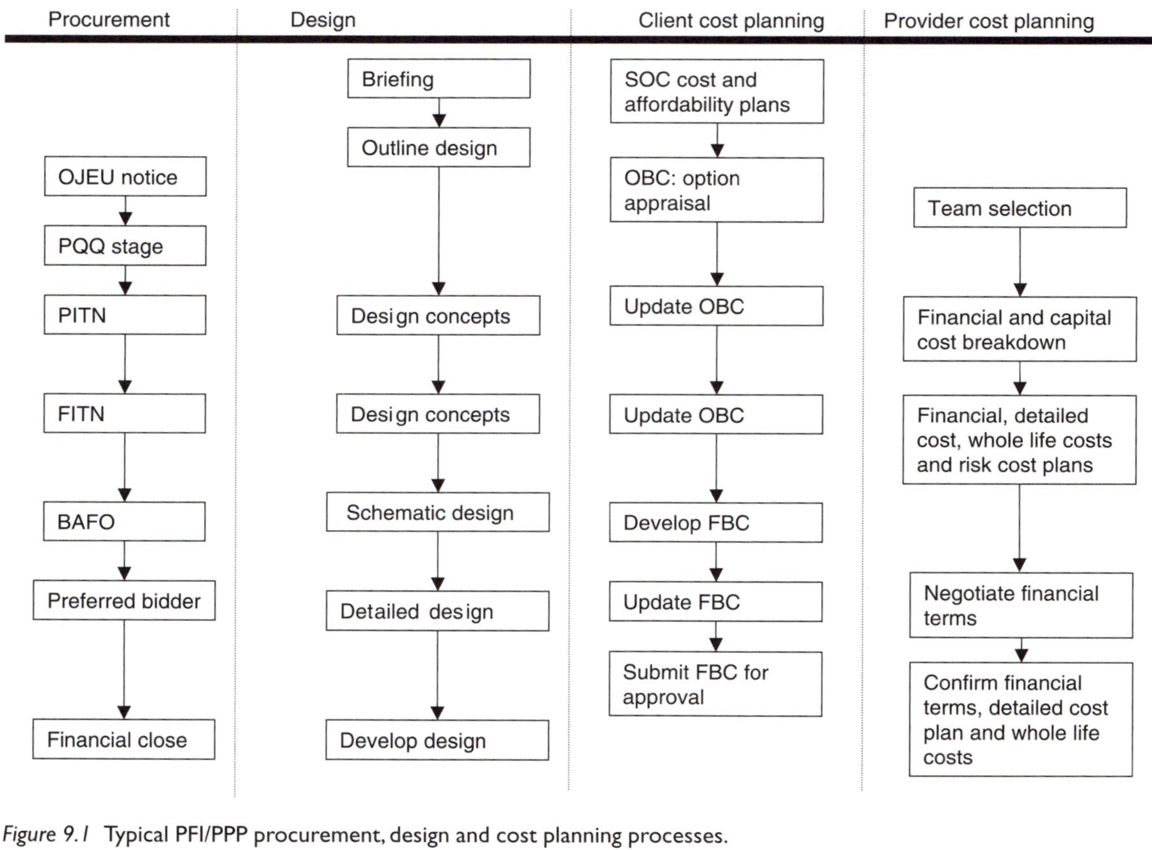

Figure 9.1 Typical PFI/PPP procurement, design and cost planning processes.

planning process begins after the pre-qualification stage. At the PITN stage, providers are invited to submit a set of financial, capital and whole life cycle cost plans.

The capital cost plans, for a health scheme, normally include:

- departmental on costs;
- electrical and lift installation;
- non-works costs;
- design and construction costs;
- engineering costs;
- preliminaries and site supervision costs;
- support cost build-ups;
- inflation allowances and risk costs.

In addition to these, the providers are required to develop and submit a separate detailed elemental cost analysis, and to provide a statement on the cost index of base cost inflation. A whole life cycle cost plan contains the following:

- a full detailed occupancy cost plan, including both soft and hard FM;
- service life of material and components;
- whole life cycle cost of elements, materials, plant and equipment;
- whole life contingency reserve plans;
- repair and replacement schedule and costs.

In addition to whole life cycle costing, the provision of financial plans and models is one of the main deviations from the traditional cost planning process. The providers may be required to develop and submit the following financial plans:

- funding structure plans including subordinated debt, mezzanine debt and senior debt. These should include information regarding drawdown schedule; repayment schedule and tenor; security required; interest rates and other fees; financial ratios and covenants; and all other revenant financial terms (see Chapter 4);
- financial model to demonstrate sensitivities in relation to interest rate changes; inflation/indexation changes; corporation tax and VAT changes; construction cost increases; life cycle cost variations; operating cost variations; payment deductions; delays in achieving financial close; construction slippage in the programme; early completion; and revenue variations;
- acceptance of the adapted standard payment mechanism.

Further details on project finance and financial terms are provided in Chapter 5.

The cost plans at the FITN stage from a provider's point of view are similar to PITN phase (but are more detailed) and are subject to a rigorous explanation and demonstration in the subsequent sections of this chapter. During the BAFO and FC stages, cost plans are refined and adjusted to any additional risk through negotiation between the procuring parties. Most of the design documentation will be finished during the FC stage. Cost plans are most likely to change during the negotiation process. At the FC phase, all contract prices, payment mechanisms, and funding terms are confirmed and frozen. The final FBC will be updated and approved by the procuring authority before the construction phase commences. A full cost and financial plan should confirm that the scheme is still within the initial budget and within the affordability parameters set at an early stage of the cost planning process. These cost plans should also demonstrate that the scheme provides value for money for both the purchaser and the provider. However, evidence from past PFI/PPP schemes shows that costestimates for PFI/PPP schemes, at most of the stages shown in Figure 9.1 have been found to be poor indicators of the outturn costs. This reinforces the view that all costs and financial terms should be reconciled with the target budget at every sign-off gateway shown in Figure 9.1. This is vital in order to keep costs under control.

Setting sign-off gateways and targets for capital and operating costs should be a critical competent of the WLC budget development at all stages of the cost planning process. Figure 9.1 demonstrates that the cost planning of PFI/PPP schemes is a progressive process. Hence, it is imperative that both the purchaser and provider should set WLC budget targets similar to Figure 9.2. This type of cost target ought to be developed at an early stage of the scheme development, based on the information set in the strategic and outline business cases. The cost targets shown in Figure 9.2 could be extracted from best performing PFI/PPP schemes. Setting capital and whole life costs in this manner will go a long way in assisting both contracting parties to control and compare the actual PFI/PPP whole life cycle budget against these benchmark targets.

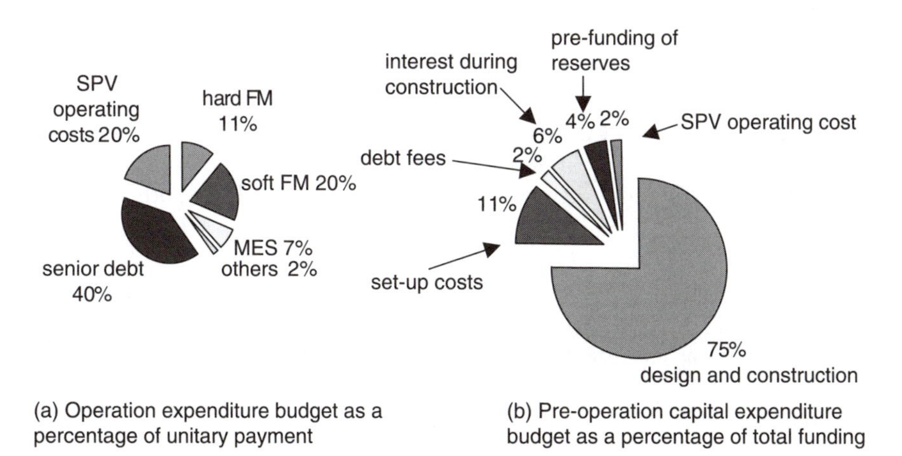

Figure 9.2 Setting capital and operating budgets targets.

Capital cost budget planning

The foundation for project cost planning has been in place since the late 1960s. Capital cost planning is concerned with the development of the PFI/PPP projects budget through the interactive processes of resource planning, cost estimating and cost budgeting. Capital cost planning here is aimed at both capital cost of the scheme as well as whole life cycle and operational costs. Cost planning starts with the establishment of the outline cost and then the cost planners work backwards to fit client requirement and accommodate current and future aspirations, whereas estimating is concerned with extracting (measuring) prices from drawings and project specifications (Ashworth 2004). Whole life capital cost planning will only be effective if clients and their advisers establish and communicate the budget limits in advance to all other stakeholders, then work in partnership with all parties to achieve the outlined capital cost plan. This approach will allow providers to determine what is possible within the prescribed capital cost plans to deliver VFM and possibly cost savings to all concerned parties. One assumes this is the case in FPI/PPP schemes, because the early involvement of the whole design, finance, construction and operation teams will lead to efficiency and cost certainty. Since at the pre-financial close stage, the design is incomplete, a target cost or outline cost plan for the scheme elements is derived, based on costs from similar past schemes. The provider's designers will have to adhere to the specified costs. This will have the advantage that the designers and providers will have a cost benchmark against which the design and specification can be compared.

Traditionally the cost plan is used as a baseline for controlling design development and provides an essential benchmark to enable projects to be delivered within budget. Theoretically, the capital cost plan is utilised to state and emphasise what is actually affordable and what the scheme should cost. Capital cost planning is based on the following two fundamental paradigms:

- the project capital cost should meet the client's brief;
- the project capital cost should accurately reflect the forecasted whole life cycle outturn costs of the proposed scheme.

The basic cost planning principles that have been advocated by Smith and Love (2004) are:

- to set a reference point for budget and cost targets for various parts of the project;
- to be used as a control method to check that the budget and cost targets are being met;
- to be used as a means for taking control action if cost targets are not kept.

The capital cost planning process is vital to control the design cost implications and simultaneously to highlight areas of risk and uncertainty and may well help clarify the stakeholders' thinking on issues of scope, specification, construction and facility operation, all of which affect the cost of building assets. Conducting capital cost risks analysis at this stage is imperative for establishing the likely outturn cost. The outcome from this procedure will assist in developing risk mitigation and response strategies. These strategies can then be employed through the whole life cycle of the facility to bring the project within the capital budget limits if the outturn cost is likely to exceed the capital cost plan. The mitigating strategies can also be utilised, rectify any defects, and, if appropriate, control the operational costs of the asset. This process will assist in the active management of the capital cost planning throughout the phases of the PFI/PPP procurement stages to ensure that costs are managed rather than just reported. One essential element of successful capital planning in PFI/PPP schemes is the strict management of change control procedures throughout all stages of the project from initial SOC to the end project concession.

If the above paradigms and principles are not followed, then problems may arise and compromises will have to be made, either in cost variations or a reduction in scope and service specifications. Both the provider and purchaser should play a key role in management of the capital planning processes, and especially in evaluating and managing the risk and uncertainty inherent in the process and the cost.

Based on the above principles, a capital cost plan is derived for a hypothetical health care facility. It is normal practice that in most cases the procurers include in their FITN documentation the requirements that the provider should deliver the following capital cost plans as part of their FITN submission:

- Capital cost breakdown: normally all bidders are requested to complete an elemental cost proforma in detail and encouraged to avoid grouping elements together;
- Detailed cost plan: separate elemental breakdowns are normally provided for each part of the construction works;
- Elemental cost summary by phase/building, if applicable;
- Capital and development costs cash flow and outturn costs.

Methods of capital costing

A wide range of capital cost estimation methods have been explored in the construction industry. Most of these methods take into account a number of significant factors that affect building capital costs, such as site conditions, contract procedures, procurement systems and market characteristics. The increasing volume of literature dedicated to the subject reflects this growing awareness. There is now much guidance on how the total costs of buildings should be modelled and a significant amount of research on cost modelling currently exists. Many of the traditional techniques thought to be in use are reported in textbooks by Ashworth (2004), Ferry and Flanagan (1991) and others. Previous studies by Skitmore and Patchell (1990), Skitmore (1991), Newton (1991), Fortune and Lees (1996), Raferty (1991), Elhag and Boussabaine (1998) reviewed all such forecasting methods and set out a catalogue of traditional as well as newer forecasting models.

Research carried out by Elhag and Boussabaine (1998) has shown that capital cost estimation of construction projects is heavily dependent upon experience and sound judgement. The respondents believe that the most popular cost estimation techniques rely on gross floor area and unit cost. Figure 9.3 portrays the responses of quantity surveyors on capital cost estimating methods used in the construction industry. Table 9.1 illustrates the accuracy of the margin of error of the most widely used capital cost estimation models. Capital cost models are classified as follows:

- *judgement*: the use of experience without qualification or quantification – experience based upon previous work would be central to such a model.
- *functional unit*: a single rate applied to the amount of provision such as £/per/KWh/m2.
- *cost per square metre area*: a single rate applied to the gross floor area (GFA) of the building. This again is widely held as a suitable method of modelling. This method fails in that it does not account for the fact that there are many other factors that are involved in determining the cost of a particular element.
- *principal item*: a single rate applied to the principal item of work such as major maintenance cost items.

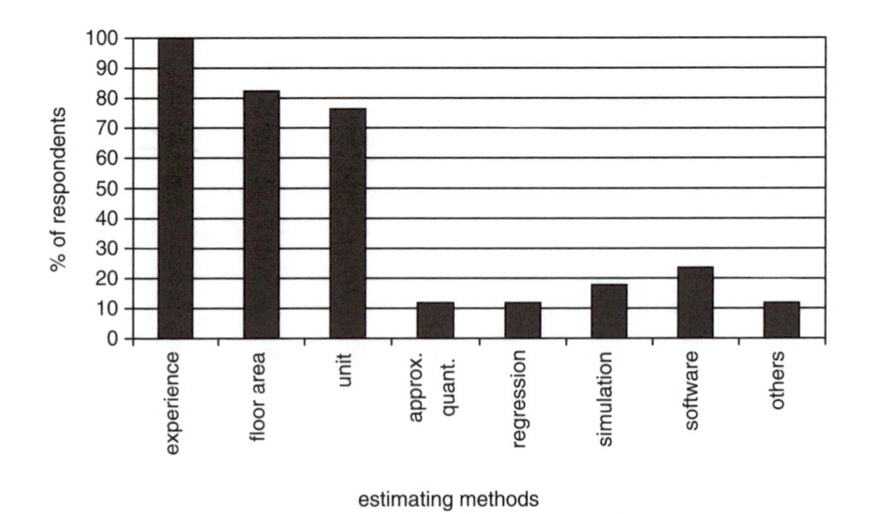

Figure 9.3 Estimation techniques (%).

Table 9.1 Accuracy of capital cost estimation methods

Estimating technique	*Contract type*	*Accuracy (%)*
Conference	Process plants	Unknown
Functional unit	All	25–30
Floor area	Buildings	20–30
Cube method	Buildings	20–45
Storey enclosure	Buildings	15–30
Approximate quantities	Buildings	15–25
Elemental analysis	Buildings	20–25
Resource analysis	All	5–8

- *interpolation method*: the interpolation of the total costs from previous projects such as historical cost data from previous hospitals.
- *elemental analysis*: the use of element unit rates to build up an estimate such as engineering maintenance unit rates.
- *significant items*: the measurement of significant items of work which are individually priced such as individually pricing major building maintenance works.
- *approximate quantities*: the measurement of a small number of items priced in groups.
- *detailed quantities*: the use of priced detailed quantities to provide a build-up of likely costs.
- *regression analysis*: cost models derived from statistical analysis of variables. The user can make estimates for a key variable using historical data from other hospitals.
- *time series models*: based on statistical analysis of trends. The user can employ statistical techniques to appraise trends in various maintenance cost centres.
- *causal cost models*: based on algebraic expression of physical dimensions. The user can express unit rates algebraically using cost coefficients enabling an estimate to be built up of total maintenance costs.
- *artificial intelligence and fuzzy theory*: the important properties of expert systems, fuzzy and other artificial techniques are invaluable in the process of modelling capital costing decisions. One should stress that the role of using these techniques in the process of capital budgeting must be supportive rather than computational.
- *net present value*: discounting future costs to present value. This is to assess whether the present value of cash inflows is greater than the present values of outflows.
- *payback method*: calculating the period over which an investment pays for itself.

- *selecting a model*: all models rely to some extent on assumptions, whether explicit or implicit. To assess the viability of a model, the user needs to have access to and understand the assumptions, which contribute to it. Unfortunately, where the assumptions are built in, it is difficult to assess and evaluate the model in question. A key issue to consider here also is that construction cost will always be uncertain. Therefore, the user must distinguish between those that include a formal measure of uncertainty (stochastic methods) and those that do not (deterministic methods).

  The selection of a model should be based on the following attributes:

  - the characteristics of data to be employed – quantitative, qualitative, large, small;
  - general knowledge of the problem to be modelled;
  - general knowledge about the boundary conditions of the model;
  - errors that the model can generate;
  - input and output targets and possible consequences;
  - understanding of accuracy, reliability, validity, confidence and sensitivity of the model to be selected;
  - understanding the parameters that define the problem to be modelled.

Capital cost breakdown

Building capital cost estimation is an experience-based process, which involves evaluations of unknown circumstances and complex relationships of cost-influencing factors. Building capital costs are affected by a large number of factors and are in the majority of cases building type related. Capital building costs are always estimated based on average conditions, then adjusted to reflect the cost of other factors such as location site accessibility, etc.

Since pricing levels vary according to location, quantity and quality factors, the price estimated in the capital cost plan should also be adjusted to account for variation caused by these factors. According to Flanagan and Tate (1997), the adjustment for price and location is reasonably straightforward and is normally computed first. The adjustment for quality, due to differences in specification, is more problematic and it is normally based on subjective judgement. In order to price the capital cost of a proposed building from an elemental cost analysis of a similar building/element type, the following expressions are used.

Adjusting for price variations:

$$Y_d = Y_a \frac{I_c}{I_a}$$

Adjusting for quantity variation:

$$Y_q = Y_d \frac{Q_c}{Q_a}$$

Adjusting for quality variation:

$Y_{qa} = Y_q \pm D_p$

Where:

Y_d = current element cost adjusted from previous projects

Y_a = element cost from previous projects

I_c = current index

I_a = index of the previous project at analysis tender date

Y_q = current element cost adjusted to quantity

Q_c = current quantity of element

Q_a = quantity of element in building analysis from previous projects

Y_{qa} = current cost adjusted for quality deferential

D_p = price deferential, due to differences in specification, as percentage of the current adjusted cost.

The above equations are in general used to adjust either the cost elements in the detailed elemental cost plan or the capital cost of the major elements in the capital cost plan. In either case, the two cost plans have to be reconcilable.

Table 9.2 shows the capital cost breakdown of a typical PFI/PPP acute health scheme comprising out-patient facilities, consulting rooms and treatment

Table 9.2 Capital cost breakdown of a typical PFI/PPP acute health scheme

Project: Acute Hospital Unit	*Element*		*Element*		*Cost/m²*	*% of total cost*
Element GFA = 5000 m²	*Quantity*	*Unit*	*Unit rate*	*Cost*	*GFA*	
Preparatory works						
Demolitions and site clearance	2000	m²	10.93	21860	4.37	0.17
Group Element Total				21860	4.37	0.17
Substructure						
Ground abnormals	500	m²	100	50000	10.00	0.39
Foundations	1500	m²	190	285000	57.00	2.20
Group Element Total				335000	67.00	2.58
Superstructure						
Frame	3500	m²	135	472500	94.50	3.64
Upper floors	3500	m²	100	350000	70.00	2.70
Roof	3800	m²	140	532000	106.40	4.10
Stairs	itm	itm		170000	34.00	1.31
External walls	3500	m²	290	1015000	203.00	7.82
Windows and external screens and doors	780	m²	595	464100	92.82	3.58

(Continued overleaf)

Table 9.2 Continued

Project: Acute Hospital Unit	*Element*		*Element*		*Cost/m²*	*% of total cost*
Element GFA = 5000 m²	*Quantity*	*Unit*	*Unit rate*	*Cost*	*GFA*	
Internal walls and partitions	3500	m²	250	875000	175.00	6.74
Internal doors	250	Nr	1500	375000	75.00	2.89
Group Element Total				4253600	850.72	32.77
Internal finishes						
Wall finishes	6700	m²	10	67000	13.40	0.52
Floor finishes	4500	m²	80	360000	72.00	2.77
Ceiling finishes	4500	m²	60	270000	54.00	2.08
Decoration	2700	m²	30	81000	16.20	0.62
Group Element Total				778000	155.60	5.99
Built-in fittings	4250	m²	165	701250	140.25	5.40
Group Element Total				701250	140.25	5.40
Services						
Sanitary appliances	250	Nr	1000	250000	50.00	1.93
Services equipment	1500	m²	65	97500	19.50	0.75
Disposal installations	250	Nr	500	125000	25.00	0.96
Water installations	1500	m²	60	90000	18.00	0.69
Heating installations	2500	m²	220	550000	110.00	4.24
Ventilation system	3500	m²	315	1102500	220.50	8.49
Electrical installations	3000	m²	220	660000	132.00	5.08
Gas installations	2	iem	20000	40000	8.00	0.31
Protection installation	1500	m²	5	7500	1.50	0.06
Medical gases, compressed air and vacuum systems	150	Nr	1800	270000	54.00	2.08
Building management system	3000	m²	40	120000	24.00	0.92
Lifts	2	Nr	179000	358000	71.60	2.76
Group Element Total				3670500	734.10	28.28
External works						
Site access roads and surface	2000	m²	70	140000	28.00	1.08
Car parks	3000	m²	60	180000	36.00	1.39
Paths and paved areas	3500	m²	50	175000	35.00	1.35
Soft landscaping	4000	m²	30	120000	24.00	0.92
Fencing and site fittings	2000	m²	20	40000	8.00	0.31
Drainage	3000	m²	125	375000	75.00	2.89
External services	2000	m²	70	140000	28.00	1.08
Group Element Total				1170000	234.00	9.01
Preliminaries and Contingencies						
Preliminaries	iem		675000	675000	135.00	5.20
Contingecies	iem		675000	675000	135.00	5.20
Design reserve/risk	iem		222271	222271	44.45	1.71
Inflation	iem		478173	478173	95.63	3.68
Group Element Total				2050444	410.09	15.80
TOTAL CONSTRUCTION COST				12980654	2596.13	100.00

rooms and care beds and operating theatres The scheme, with a gross internal floor area of 5000 m^2, comprises three storeys of accommodation. The plant room is situated at the roof level. To assist in the development of the capital cost breakdown of this scheme, the capital budget is analysed into the following group elements, and the capital cost of building is normally subdivided into the following cost groups.

Building substructure

The capital cost of substructure depends to a large extent on the type of soil, the site conditions and the foundation type. It is obvious that different type of foundations have different effects on the final capital cost of buildings. The average cost £/m^2 for projects that uses strip and pile foundations is 6 per cent and 26 per cent higher than the normal average. On the other hand, the average cost £/m^2 of pad foundation projects is 27 per cent lower than the normal average. The choice of foundation type is normally based on the bearing capacity of the site soil. Studies showed that the average cost £/m^2 for sites that have good and moderate soils is 24 per cent and 39 per cent lower than the normal average of cost per m^2 (within the same type of buildings), whereas the average cost £/m^2 of site with poor soil conditions is 36 per cent higher than the average cost (Boussabaine and Elhas 1999a).

The percentage of capital spent on substructure depends largely on the specification of the project under estimation, for example, complex building sites that need reinforced concrete piling, reinforced concrete slab to basement, retaining walls, formation of lift pits and all other work that is associated with basement construction. The percentage capital spent on this type of substructure probably will exceed 6 per cent of the total construction cost. But, in simple framed buildings that require only reinforced *in situ* concrete column bases and ground-floor slab with edge beams, lift pits and associated ground works, the percentage of capital cost of this type of substructure might be lower than 5 per cent of the total building cost.

Since the soil-bearing capacity is not adequate to support the superstructure load, reinforced concrete piling, reinforced concrete slab to basement, retaining walls, formation of lift pits and all other work that is associated with basement construction are specified. To estimate the price of the substructure for the current scheme, projects that have comparable types of foundation and soil conditions are analysed. The analysis shows that the average cost for similar substructures are priced at £110 per m^2 GFA. This price will need to be updated for inflation (no adjustment for location differences is required, due to the fact that data are extracted from projects within the same region):

$$\text{Revised cost per m}^2\text{ GFA} = 110 \times 191/169 = £124$$

$$\text{Elemental cost} = 5000 \times 124 = £620000$$

Superstructure

In general, the superstructure consists of the frame, upper floors, roof, stairs, external walls, internal walls/partitions and internal doors. Previous studies have demonstrated that reinforced concrete frames are the most expensive, followed by steel structures, and the least expensive are the load-bearing buildings. This might be attributed to several variables such as type of materials, formwork and plants required, and also the time required for the erection of different types of frames. The average cost £/m^2 for load bearing buildings is 22 per cent lower than the total average capital cost of buildings. On the other hand, the average cost £/m^2 of steel and concrete projects are respectively 0.6 per cent and 13 per cent higher than the total average (Boussabaine and Elhag 1999a). On average, the concrete frames at over £150/m^2 are a bit higher that the steel frames at over £90/m^2. Of course, rates must always be subject to adjustment to account for specification and any other design and contractual factors. It is also suggested that the preliminaries on most concrete-framed facilities tended to be marginally higher than on the steel-framed schemes. An allowance of 5 per cent or over of building capital cost probably is sufficient for a structural steel frame with standard and cellular beam sections. For reinforced concrete frames, this percentage depends largely on the grid size, type and thickness of the slab and the presence of *in situ* concrete sheer walls. In any case, it is expect to be higher than the steel option; in some cases it could be twice as high as the steel frame. However, there is some recent evidence to show that the cost of steel raw materials is increasing rapidly. If this trend continues, this may tip the balance in favour of concrete structures. This will also have an effect on the cost of concrete steel reinforcement. One should also remember that the selection of building frames is dependent on other factors.

To derive a cost for the frame specified in Table 9.2, we assume that the current scheme has the same number of floors and storey heights. The frame specified is of standard structural steel frame with standard cellular beam sections. Similar analysed projects show that comparable frames are priced at £95 per m^2 GFA, including fire protection and all necessary decorative finishes. The data from the analysis have been updated for inflation and regional variation as follows:

$$\text{Updated cost} = 95 \times 191/169 = 107$$

$$\text{Element cost} = 5000 \times 107 = £535000$$

Upper floors

The cost of the upper floors element varies according to the construction technology adopted and design factors such as length of span, loadings, thickness of the slab, etc. In the proposed scheme, it is envisaged that a composite contraction method is going to be used, based on cellular beams, metal deck, reinforced concrete slab and power floating deck. Cost analysis shows that the average cost of comparable upper floors elements is £70 per m^2 GFA. That total cost of this element is calculated and updated for inflation as follows:

Revised cost = £70 × 191/169 = 79

Element cost = 5000 × 79 = 395000

Roof

The roof typically represents approximately over 2–20 per cent of the total building cost, depending on the number of storeys, roof type, roof structure, roof coverings and roof drainage. For example, for a single storey building with slated pitched and conical roof, the element cost would be very significant and may account for up 15 per cent of the total building cost. The average roof price is dependent to some extent on building type. Warehouses and factories have the lowest cost per metre square of roof area, whereas offices and hospitals have the highest cost per metre square of roof area.

Since the case study building has a roof level plant room, a flat composite construction roof is specified. The roof is constructed from a thick reinforced concrete slab on a metal deck. Associated roof works includes power floating deck, bitumen prime, insulation, felt vapour barriers, rainwater pipes and gutters, etc. The cost analysis of comparable roofs shows that the cost is approximately £90 per m^2 GFA.

External walls

The external envelope comprises external walls, cladding, windows and external doors. There are differences in the way in which specifications and costs of these element decisions are made under a particular type of contract. The element specification and costing in PFI/PPP schemes are similar to design and build, where the provider takes responsibility for the design, construction as well as the operation, and the provider is widely recognised as having a greater influence on the element specification and cost. Normally, the majority of initial element specification decisions are taken by the architect (in PFI, designers are responsible to the provider), however, the provider often suggests changes to the architect specification, to develop low-cost, low-maintenance and low-risk solutions. The client has little or no input in specifying the external envelope in PFI/PPP schemes.

The cost of the external envelope is very significant and can represent up to 25 per cent of the total building cost. Obviously, the capital cost of this element is dependent on several pricing factors, such as the type of construction technology used, the services zone requirement, the complexity of the building shape, the building height, the ratio of wall to openings, etc. Hence, the importance of minimising the floor-to-floor height to minimise the capital cost of cladding and external walls. This also has a direct relationship with the size of the services zone. The rule of thumb is the thinner the overall structural and services zone, the cheaper the cladding and external walls.

The diversity and complexity of existing external envelope construction technology make it very difficult for a cost estimator to derive an accurate price per square metre ground floor area (GFA) for this type of element. Probably, it is best if the analyst uses data from similar schemes with particular emphasis on the complexity of the shape and size of the building.

The pricing of windows and external doors varies considerably, depending on the type and function of building. For example, where fire safety is an issue, one expects the cost of these elements to increase moderately. Probably the best way of extracting the price of these type of elements is to directly acquire quotations from sub-contractors and manufacturers.

For the purpose of developing the capital cost of the case study, it assumed that the external envelope will consume 15 per cent of the total building capital cost. The assumption is based on the specification of an aluminium curtain walling, composite cladding panels, secondary steel support and sealed double glazed units. All external doors and windows are aluminium framed. The price of such specification is estimated at £200 per square metre GFA.

Internal finishes

Internal finishes consists of internal walls/partitions, wall finishes, internal doors, floor finishes, ceiling finishes and furniture and fitting cost group elements. Of these element costs, probably the furniture and fitting is the most costly. The cost of fitting out and removal of internal fittings is largely dependent on the layout and construction technology used to separate spaces. Most common layouts include open plan, cellular accommodation or a combination of both. It is envisaged that with the development of innovative materials and construction methods, such as mobile wall partitions, the cost of set-up and clearing will be reduced dramatically. Also fitting out costs depend on the quality and standard of fittings. The cost of medical equipment depends on the type of equipment to be supplied and installed (fixed, loose furniture, nurse stations). Also it depends on space to be equipped and finished. For example, dentistry rooms are more costly to fit out. Capital cost spends on furniture and fittings could well be over 5 per cent of the total building cost. Internal walls and partitions also consume around 5 per cent or more of the building capital cost. The most costly items in wall finishes are ceramic wall tiling and aseptic sterile proprietary to cover surface finishes to clinical and procedure areas. The major other areas are painted with emulsion. So the percentage of capital cost of wall finishes could be as high as 5 per cent, depending on the specification. There is little variation of floor finishes for health schemes; hence an allowance of up to 4 per cent of the building cost will be sufficient. Both internal doors and ceiling are normally specified to comply with fire safety regulations. Their capital cost is around 3 and 2 per cent of the total building cost respectively.

Engineering

Building services costs contribute significantly to the total cost of projects. It is well documented in the literature that the cost of buildings increases with the complexity of the building services. With mechanical engineering installations becoming ever more sophisticated and complicated, their capital and operational costs are becoming increasingly significant. Research carried out by the author showed that the average cost £/m^2 for projects with moderately and complex services are 6 per cent and 26 per cent higher than the total average of the sample of the project investigated, whereas the average £/m^2 for projects with easy services is lower by 55 per cent than total average. The percentage variation from the total mean is 45 per cent (Boussabaine and Elhag 1999a). These results reinforce the view that building services complexity has a direct effect on the cost of projects.

The cost elements that are normally included in a typical FITN document are shown in Table 9.3. Ventilation and air treatment systems in health schemes constitute a major capital investment as shown by their percentage of the total construction costs. Capital costing from previous similar scheme projects shows that the ventilation and air treatment systems may constitute up 10 per cent of

Table 9.3 Cost elements included in a typical FITN document

Project: Acute hospital Gross floor Area in m^2: 5000	*£'000 Apr 06 MIPS 449*	*£'000 Outturn MIPS 469*	*£'000 % of total MIPS 469*
Substructure			
Abnormals	0		
Demolitions and site clearance	50	52	0.31
Foundations	285	298	1.75
Element Total	335	350	2.05
Alteration works			
Alterations	0		
Repairs to existing fabric	0		
Element Total	0		
Superstructure			
Frame	473	494	2.90
Upper floors	350	366	2.15
Roof	532	556	3.26
Stairs	170	178	1.04
External walls	1015	1060	6.22
Windows and external screens and doors	464	485	2.84

(Continued overleaf)

Table 9.3 Continued

Project: Acute hospital *Gross floor* *Area in m^2: 5000*	*£'000* *Apr 06* *MIPS 449*	*£'000* *Outturn* *MIPS 469*	*£'000* *% of total* *MIPS 469*
Internal walls and partitions	875	914	5.36
Internal doors	375	392	2.30
Element Total	4254	4443	26.07
Internal finishes			
Wall finishes	67	70	0.41
Floor finishes	360	376	2.21
Ceiling finishes	270	282	1.65
Decoration	81	85	0.50
Element Total	778	813	4.77
Furniture, fittings and equipment			
Furniture and fittings	401	419	2.46
Workshop and other FM equipment	150	157	0.92
Other services equipment	150	157	0.92
Other equipment	0	0	0.00
Element Total	701	732	4.30
Services			
Sanitary appliances	250	261	1.53
Services equipment	98	102	0.60
Disposal installations	125	131	0.77
Hot and cold water installations	90	94	0.55
Heating installations	550	574	3.37
Ventilation system	1102	1151	6.75
Electrical installations	660	689	4.05
Gas installations	40	42	0.25
Fire protection	8	8	0.05
Medical gases, compressed air and vacuum systems	270	282	1.65
Communication installations	120	125	0.74
Lifts	358	374	2.19
Builders' work in connection with engineering services	90	94	0.55
Element Total	3761	3929	23.05
External works			
Site clearance and demolition	25	26	0.15
Preparatory earthworks	0	0	0.00
Site access roads and surface car parking	140	146	0.86
Car parks	180	188	1.10
Paths and paved areas	175	183	1.07
Soft landscaping	120	125	0.74
Fencing and site fittings	40	42	0.25
Other site works (retaining walls, etc.)	0	0	0.00
Drainage	375	392	2.30
External services	140	146	0.86

External lighting	5	5	0.03
Buildings work in connection with external services	100	104	0.61
Minor building works	25	26	0.15
Element Total	1325	1384	8.12
Alteration works			
Alteration/Diversions works	0		
Temporary structures and other enabling works	0		
Element Total	0		
Preliminaries			
On-site staff	135	141	0.83
Sundry costs	80	84	0.49
On-site labour	95	99	0.58
Materials and goods	90	94	0.55
Scaffolding	110	115	0.67
Plant, consumables, stores and services	95	99	0.58
Health and safety	70	73	0.43
Insurances	0	0	0.00
Element Total	675	705	4.14
Contingency			
Design reserve	222	232	1.36
Construction contingency	675	705	4.14
Element Total	897	937	5.50
Design and technical fees			
Architects	100	104	0.61
Structural/electrical	90	94	0.55
Quantity surveyors	75	78	0.46
Planning supervisors	90	94	0.55
Healthcare planners	80	84	0.49
On-site supervision	120	125	0.74
Element Total	555	580	3.40
Non-works costs			
Planning fees	10	10	0.06
Building regulations	10	10	0.06
Removals/decanting costs	0	0	0.00
Site surveys and investigations	15	16	0.09
Element Total	35	37	0.21
Risk allowance			
Risk allowance	3000	3134	18.39
Element Total	3000	3134	18.39
Inflation			
To financial close	100	104	0.61
During construction period	478	499	2.93
Element Total	578	604	3.54
TOTAL CAPITAL COSTS	16316	17043	100.00

the building total cost and at a cost of over £200/m^2. Electrical installation takes up a considerable proportion of the building capital cost (more that 5 per cent of building total cost) and costs an average of £120/m^2. The next most expensive item in the engineering cost group is lift installations. The cost depends on the type of lift and number of floors to be serviced by the lift. An allowance of 3 per cent of the building cost for each lift is sufficient. All modern building facilities, especially healthcare schemes, need sophisticated and extensive communication installations. Normally communication installation might account for up to 4 per cent of the total building cost, whereas an allowance of up 3 per cent of the building cost will be sufficient to cover costs for specialist installation, such as medical gas systems.

External works

Normally the elemental cost analysis reported in cost models and data bases is only for building. The cost of roads, landscaping, site enclosure and any services run are usually estimated separately under the heading of external works. It is expected that the cost of external work varies according to building types, site, etc. Generally it is assumed that large capital schemes, cramped sites and multi-storey construction tend to have low percentage additions for external works. For health schemes, this percentage ranges from 10–25 per cent of building cost. Probably an average allowance of 15 per cent for hospitals and 20 per cent for health centres is sufficient to cover the external works capital expenditure.

Alteration works

In many PFI schemes, alteration and updating current building stocks are part of the contract. If this is the case, then the providers are obliged to submit details and pricing of such work. The cost of altering existing building stock can vary between 0.5 and 0.75 per cent of the new build, depending on the condition of the existing stock and the specification requirements.

Non-works costs

Most of the sites require further enabling works before the actual construction process starts. These costs are related to issues such as demolition, decontamination of site planning fees, building regulations, statutory or local authority charges, removals/decanting costs, site surveys and investigations, etc. For the purpose of the case study, an allowance of 1 per cent of works cost is made for this purpose. The cost of these items is normally based on quotations from subcontractors or rates from similar building activities.

Preliminaries

The majority of the preliminaries' cost is time-related. As a general rule, the cost of the provider's (or their contractor's) preliminaries is related to the issue of site management, site establishment, testing and commissioning of building services installations and on-site facilities. The majority of preliminaries are time-related and are influenced by the contract period site characteristics. Sites that are located in inner cities and confined spaces may require special or extra resources. These sites will incur a higher figure for preliminaries. For example, research by the author has showed that the average cost £/m^2 of facilities build on highly restricted sites is found to be over 50 per cent higher than the normal average cost per m^2. Site topography factor, i.e., levelled sites, gentle sloping landscapes, and steep sloping sites, also have an impact on the cost of both substructure and superstructure. It is expected that the cost per m^2 of levelled and gently sloping sites is lower than the average. In contrast, the average cost £/m^2 of a steep sloping site is expected to be over 20 per cent higher than the normal average. It is recognised that contaminated soil will increase the project cost if all other factors remain the same. On average, the cost £/m^2 of buildings on contaminated sites is 23 per cent higher than the normal average (Boussabaine and Elhag 1999a). The availability of working space is important in estimating the capital cost. Research has demonstrated that projects that are built on highly restricted working spaces have a higher cost per m^2.

There are several ways of documenting preliminaries' cost. The most usual method is to show the cost of preliminaries as a lump sum. Normally the value of preliminaries might account for approximately 10–20 per cent of the total building cost depending on the type, size and length of the construction duration and site characteristics. In the present case study an allowance of 15 per cent is estimated to cover the preliminaries costs.

Contingencies

As discussed in Chapter 7, the contingency value is an *ad hoc* value decided by the purchaser during the development of OBC. However, most of the clients will instruct the bidders to include an amount in their FITN submission to cover and absorb any unforeseen costs. The value allocated to contingencies should have little effect on the pricing of the FITN documentation. The value of the contractor's contingencies often represents approximately up to 5 per cent of the contract sum. Readers can refer to Chapter 7 for more details.

Design and technical fees

The level of fees is largely dependent on the complexity of the PFI deal. At the FITN stage, bidders are required to specify and justify the proportion of professional fees to be treated as revenue expenditure for tax purposes. Bidders should also specify and justify the proportion of professional fees to be treated as qualifying expenditure for capital allowances purposes (if applicable). Professional fees are derived for the following services:

- structural engineers
- electrical engineers
- quantity surveyors
- performance bond
- project management
- project sponsorship
- independent certifier
- legal fees
- site supervision and inspection
- bank due diligence fee
- technical fees
- financial fees
- bank arrangement fees.

Risk allowance

Part of the FITN stage is that submission bidders are required to put a value on risk allowances for the construction works that may be outside the contract sum but within the overall project budget. The risk allowance is normally assessed for identified risks. It is important that bidders must ensure that the risk allowance for each element of the project is reasonable and up to date (the risk allowance should be for identified risks only and not an assumed contingency provision). A thorough risk allowance, based on extensive risk analysis, will lead to an increase in the base estimate of capital costs of the scheme and accordingly it will result in a lower risk allowance allocation. The risk allowance is the sum which is allocated to cover the expected costs of all risks identified by the bidders, including the project sponsor, acting collaboratively and participating as stakeholders in the success of the project. The value of risk allowance is estimated from the risk register, taking into account the probability of each risk occurring, and potential impact. Hence, deriving a value for risk allowance is project-dependent. However, this risk allowance, from the provider point of view, should at least account for the difference between the PFI/PPP bid and the public sector comparator. The reader can consult Chapters 12 and 13 for further details.

Detailed cost plans

Detailed elemental cost analysis showing the capital cost of construction, including all fixed furniture, fittings and equipment and IT infrastructure is developed at this stage. Bidders are in general requested to provide a detailed and fully quantified cost plan, which arithmetically supports the figures in within the capital cost breakdown shown in Table 9.1. In all circumstances, the detailed elemental cost analyses must reconcile with the capital cost plan and financial model. As illustrated in Table 9.1, the cost plan is presented in sufficient detail to identify the specification, quantity and rate for all major materials and components adopted for both the builder's work and the engineering installations. It is advisable that the cost analyst should avoid grouping elements together. In some PFI/PPP cases, the detailed cost plan, required by the client, may include the following specification:

- A detailed and fully quantified breakdown of preliminaries, which arithmetically supports the figures, split into the categories, identified in Table 9.1. In some cases this may necessitate the provision of schedules that identify and cost numbers of staff in terms of man days, confirm provision for site accommodation, cranage, scaffolding, skips and other associated site management costs.
- Details to support the contingency and risk allowance assumptions and computation.
- A detailed breakdown of assumptions and calculations used to account for inflation to financial close and inflation during the construction period. All indices used for their purpose should be identified.

The format for presenting the detailed cost plan varies according to project complexity and the purchaser specification. A typical example of how to present a detailed elemental cost analysis is shown in Table 9.3.

Elemental cost summary

It is good practice for the cost analyst to provide a concise elemental cost summary. From the information set out in Tables 9.2 and 9.3 the cost analyst should be able to summarise the capital costs in the form illustrated in Table 9.4 or other similar format. The summary is normally based on a predefined MIPS by the client and at MPIS at outturn specified by the bidder. The outturn MPIS is to take into consideration price inflation during the negotiation of the contract and construction period. As shown in Table 9.4, the summary must cover and identify separately the whole of the anticipated capital expenditure on the project including:

- building and engineering costs, broken down into individual phases and/or buildings, as described in Tables 9.2 and 9.3;
- external works and alterations;
- preliminaries and contingencies;
- contractors' fee;
- design and technical fees;

Table 9.4 Elemental cost summary

Project: Acute Hospital Unit *GFA = 5000 m²*	*£'000* *Apr-06* *MIPS 449*	*£'000* *Outturn* *MIPS 469*	*£'000* *% of Total cost at 2*
Element	*1*	*2*	*3*
Building and engineering	9829	10267	55.87
External Works	1325	1384	7.53
Alterations	0	0	0.00
Preliminaries	675	705	3.84
Contingencies	897	937	5.10
TOTAL CONSTRUCTION COSTS	12726	13293	72.33
Contractors' Fee	700	731	3.98
Design and Technical Fees	555	580	3.15
Non-works Costs	35	37	0.20
Risk	3000	3134	17.05
Inflation	578	604	3.29
TOTAL CAPITAL COSTS	17594	18378	100.00

- non-works costs;
- risk;
- inflation.

Capital and development costs cash flow

Cash flows for construction works as detailed in the elemental cost analysis including all fixed equipment items, professional fees and statutory charges are computed and presented in Tables 9.2 to 9.4. Normally, bidders are required to present cash flows of projected capital spend on a monthly basis in the format shown in Table 9.5. Cash flows presented in Table 9.5 are developed based on the construction programme and are project depended. The reader can consult Chapter 6 for in-depth analysis of cash flow in the PFI/PPP procurement method.

Estimating operating cost budget

The capital investment normally represents only a fraction of the operation costs over a project's life cycle. Operational costs are attributed to all revenues associated with running a building facility and include:

- catering services
- waste management services
- portering services
- postal services
- car parking services

Table 9.5 Development cash flow

Project: Acute Hospital Unit	*Total*	*Monthly periods*															
GFA = 5000 m²	*MIPS 469*	*1*	*2*	*3*	*4*	*5*	*6*	*7*	*8*	*9*	*10*	*11*	*12*	*13*	*14*	*15*	*16*
Element	*£'000*	*£'000*	*£'000*	*£'000*	*£'000*	*£'000*	*£'000*	*£'000*	*£'000*	*£'000*	*£'000*	*£'000*	*£'000*	*£'000*	*£'000*	*£'000*	*£'000*
Building and engineering	10267	52	149	149	165	165	165	400	450	500	525	967	1100	1200	1300	1350	1630
External works	1384										50	75	100	220	250	275	414
Alterations	0																
Preliminaries	705			20	20	25	30	40	55	55	60	65	65	65	70	70	65
Contingencies	937			50	50	50	60	60	70	70	80	80	80	87	90	90	70
CONSTRUCTION COSTS	13293																
Contractors' fee	731																
Design and technical fees	580	70	60	50	50	60	40	40	40	30	20	20	20	20	20	20	20
Non-works costs	37	10	10	17													
Risk	3134			drawdown as required													
Inflation	604	52	52	33	20	33	37	35	35	32	33	32	40	60	55	42	65
TOTAL CAPITAL COSTS	18378																

- reception services
- cleaning services
- security services
- helpdesk services
- building maintenance
- asset renewals.

These costs are in general divided into hard and soft FM. Soft FM includes cleaning, security, portering, postal services, utilities and rates. Hard FM includes maintenance, repair and replacement of components. This section will discuss the issues involved in pricing soft FM provision. The analysis and discussion will be based on actual data from a sample of acute teaching hospitals outside the London area. All data are self-reported by the NHS trusts and recorded in a database known as ERIC (Estates Returns Information Collection System). Table 9.6 is composed of descriptive statistics that include means and standard deviations of costs. The data sample shown in Table 9.6 is quite varied. For example, the standard deviations exceed mean cost of postal services over occupied floor area (OKA) for input attributes of studied hospitals. The range of all the variables is large. This shows the complexity and difficulty of deriving such costs for new schemes. In the following subsections the cost of some of the soft FM elements are extracted and presented in terms of cost per occupied floor area (OFA).

Energy costs

Energy costs are the total cost of electricity, gas, coal, oil and any other source of energy, per occupied floor area. The annual cost of energy in this sample of hospitals ranges from £9 up to £138 per OFA with a mean value of £38 per OFA. The energy cost here depends on the size, the age of the building and the type of fuel used to generate the heat. For large schemes like hospitals,

Table 9.6 Operational facilities data from a sample of acute hospitals

Input/Output	*Mean*	*SD*	*Minimum*	*Maximum*
Total energy costs/OFA	38.09	31.73	9.28	132.01
Total electrical energy cost/OFA	16.53	11.71	4.96	43.97
Total water and sewage cost/OFA	6.58	4.32	1.64	15.55
Total cost of waste/OFA	6.22	4.24	0.56	14.14
Total engineering maintenance cost/OFA	38.87	28.26	5.79	117.13
Total building maintenance cost/OFA	17.10	14.51	4.46	51.48
Total cost of cleaning/OFA	62.43	51.70	16.71	207.19
Cost of laundry and linen services/OFA	16.51	9.52	6.62	35.70
Cost of security services/OFA	5.12	2.93	2.18	11.72
Total cost of portering services/OFA	27.09	20.06	5.76	80.71
Total cost of postal services/OFA	6.94	7.10	1.86	30.53

the following options should be considered in developing the capital and operational costs of the energy systems:

- centralised or decentralised boiler houses;
- choice of fuels;
- medium for transport of heat around each building, whether water or electrical;
- electrical power distribution and generation;
- low pressure hot water heating and domestic hot and cold water distribution;
- ventilation philosophies including flexibility, general accommodation and specialist departments;
- ceiling panel versus low surface temperature radiator space heating;
- local electrical hot water versus central gas-fired;
- combined heat and power.

Electrical energy

This cost centre refers to the total annual cost of electricity used including energy costs associated with feeding CHP plant together with energy used to service building assets. Statistical analysis of this cost shows that the electricity costs varies from £4 up £43 per OFA as shown in Table 9.6. The mean value is £16 per FAO. This huge variation in the cost of electricity is mainly attributed to the size of the facility to be serviced. Opportunities to minimise disruption to the client associated with the procurer's planned maintenance activities must be carefully considered and cost-effective solutions should be identified and priced at the FITN stage. Another important aspect of electrical supply to hospitals is to provide standby portable power generation systems to power the critical health system in the unlikely event of supply failure. Aspects of flexibility for future growth and service resilience should also be priced.

Water and sewage cost

This cost element relates to the total cost of metered water and sewage. This also should include water costs associated with on-site central processing unit(s) (e.g. laundry) and all water and sewage costs associated with non-metered premises and surface water highways and drainage charges. Analysis has demonstrated that the annual mean average value of this cost centre is £6 per OFA. The actual cost fluctuates between £1 and £15 per OFA.

Cost of waste

This cost element embodies the total cost of disposal of all domestic, special and clinical waste generated from building occupancy, including the cost of transport and disposal. The cost per OFA of this element is similar to the cost of water and sewage as shown in Table 9.6.

Cost of cleaning

The cost of cleaning encompasses the total pay and non-pay cost of cleaning services. This should include all elements relating to the cost of employment of directly employed staff for the organisation site, inclusive of managers, supervisory and administrative staff. It should also include the cost of contract staff, fees, materials, equipment provisions, and uniform costs. This cost element is significant and is shown in the analysis of current expenditure in Table 9.6. The cost per OFA of this element varies from £16 up to £200 per OFA with a mean value of £62 per OFA. For acute health facilities, this element should be priced towards the upper limit. The high cost of this element could be attributed to the fact that cleaning activities are mainly labour-intensive and the frequency of cleaning in health buildings is very high.

Cost of laundry and linen services

This cost element is related to the provision of laundry and linen services. The analyst should include in the pricing of this element all fees, labour, materials, consumables, associated with directly employed, contracted out staff, disposables, transport, collection and distribution to the point of use, and replacement and repairs. The cost of this element is directly related to number of bids and inpatient days. The higher these attributes, the higher the price for this element ought to be. From the range of actual cost of this element shown in Table 9.6, it appears that we should expect the pricing of this element to be in the range of £16 up to £35 per OFA.

Cost of security services

The analyst should aim to price the total pay and non-pay cost for the provision of all security services associated with the activities and operation of the facility in question. The price should include all costs associated with equipment, uniforms, consumables and labour for directly employed and contract staff designated to security duties. System maintenance costs, e.g. CCTV and door access, ought to be priced under the maintenance budget. Cost studies have demonstrated that the analyst should anticipate the price of this element to be in the range of £3–12 per OFA.

Cost of portering services

The pricing of this element is related to pay and non-pay cost for the provision of all portering services that are envisaged for a particular building facility.

The pricing should take into consideration all elements relating to the cost of directly employed and contract staff, fees, materials, training, equipment provisions, and uniform costs for the purpose of carrying out portering duties and courier services. With regard to potering services, cost studies show that the price of this element is expected to be in the range of £6 to £80 per OFA and with a mean value of over £27 per OFA.

Cost of postal services

The pricing of post services involves determining the total pay and non-pay cost for the provision of postage services. This should include all directly employed and contract labour, postage and courier charges, equipment provision, uniforms, consumables, stamps, sorting and distribution costs for delivering incoming mail and for the collection of outgoing mail inclusive of letters, packages and parcel post. It should include all costs associated with the internal postal service and, where appropriate, equipment maintenance. It is surprising to see the actual cost analysis shows that the mean average cost of this element is higher than some of the essential services, see Table 9.6. The analyst should price this element in the range shown in Table 9.6.

Based on the above analysis, the prices for soft and hard facility management shown in Table 9.7 are derived.

Estimating repairs and maintenance cost budget

Maintenance costs are related to the expenditure of keeping a building facility in good repair and working condition. The cost for maintaining and operating a building can far outweigh the initial cost, and contributes significantly to the total cost of a building ownership. Hence, the importance of having a long-term maintenance and repair strategy to maximise efficient use of the building and keep maintenance and running costs as low as possible.

Maintenance work can be divided into two main areas: planned (preventative maintenance) and unplanned (reactive maintenance). Planned maintenance can be further divided into two more subdivisions. Schedule-based, which is maintenance carried out on a cyclical basis before failure, irrespective of the condition of the elements, and condition-based, in which maintenance is carried out when an element has reached the end of its life span, usually as a result of surveys or tests. Unplanned maintenance can also be further divided into routine maintenance which is the repair or replacement of defective items and emergency maintenance, which is maintenance necessary to avoid unavailability consequences, especially in PFI/PPP schemes, which has an impact on the payment by the client to the provider.

In developing the price of repairs, replacement and maintenance at FITN stage, the cost analyst is confronted with the fact that the design is still incomplete, which make it very difficult for him to extract the service life of

Table 9.7 An example of pricing soft and hard FM

Project: Acute Hospital Unit	*Annual periods*														
GFA = 5000 m^2 *Element*	*1 £'000*	*2 £'000*	*3 £'000*	*4 £'000*	*5 £'000*	*6 £'000*	*7 £'000*	*8 £'000*	*9 £'000*	*10 £'000*	*11 £'000*	*12 £'000*	*13 £'000*	*14 £'000*	*15 £'000*
Energy	950	1235	1606	2087	2713	3527	4585	5961	7749	10074	13097	17026	22133	28773	37405
Electrical	975	1268	1648	2142	2785	3620	4706	6118	7953	10339	13441	17474	22716	29530	38389
Water and sewage	1490	1937	2518	3274	4256	5532	7192	9350	12154	15801	20541	26703	34714	45128	58667
Cost of waste	105	137	177	231	300	390	507	659	857	1113	1448	1882	2446	3180	4134
Engineering maintenance	85	111	144	187	243	316	410	533	693	901	1172	1523	1980	2574	3347
Building maintenance	11	14	19	24	31	41	53	69	90	117	152	197	256	333	433
Cost of cleaning	145	189	245	319	414	538	700	910	1183	1538	1999	2599	3378	4392	5709
Laundry and linen services	746	970	1261	1639	2131	2770	3601	4681	6085	7911	10284	13370	17380	22594	29373
Cost of security services	665	865	1124	1461	1899	2469	3210	4173	5425	7052	9168	11918	15493	20141	26184
Cost of portering services	195	254	330	428	557	724	941	1224	1591	2068	2688	3495	4543	5906	7678
Cost of postal services	107	139	181	235	306	397	516	671	873	1135	1475	1918	2493	3241	4213
Supplies and services	900	1170	1521	1977	2570	3342	4344	5647	7342	9544	12407	16129	20968	27259	35436
Administration costs of contractor	500	650	845	1099	1428	1856	2413	3137	4079	5302	6893	8961	11649	15144	19687
TOTAL	6874	8936	11617	15102	19633	25523	33179	43133	56073	72895	94764	123193	160151	208196	270655

components. Without knowing the exact specifications of building components, it is very hard to estimate their replacement cycles. Hence, the first set-up in defining maintenance and replacement costs is to predict replacement cycles or service lives for each component in order to price their life cycle replacement and periodical maintenance costs.

All building assets deteriorate over time. Designing these assets and their components for a particular service life and maintaining them in a safe condition during their entire service life and during the concession period and beyond is a very critical issue in PFI/PPP procurement method. The design specifications plus the design life of the assets will significantly affect the replacement frequency of components. Most of buildings are expected to last over 60 years. Table 9.8 shows the life expectancy of several building elements. This type of information can be extracted from several sources. But the final decision in estimating service life of components should be based on judgement of the in-use condition of the component. As can be seen in Table 9.8, critical elements are expected to last the life of the building; these should include components that are difficult to replace such as substructure, structure, etc. But components such as windows, internal fixtures, roof coverings, and mechanical and electrical services will need maintaining and replacing several times during the contract concession period. In developing the service life shown in Table 9.8, the analyst should bear in mind that a number of factors can affect the life expectancy of a component. The issues to be considered include:

- the rate and quality of maintenance during service;
- rate of use, i.e., wear and tear;
- environmental conditions;
- warranties;
- availability of spare parts and components;
- consumables in the expected lifespan;
- availability of skilled and approved maintenance subcontractors;
- quality of the component;
- quality of installation or construction.

Several mechanisms are currently in place that can assist professionals to make component and building service life predictions. Generally, methods of predicting the remaining life of building components can be classified into two distinct categories:

- deterministic prediction methods;
- stochastic prediction methods.

The use and selection of a particular method depend on factors such as availability of data, complexity of the analysis, etc. Readers can consult Boussabaine and Kirkham (2004) for an in-depth analysis of the issues involved.

Having identified the service life of components and their replacement cycles, the analyst now needs to decide on types of maintenance to be considered for the building element. In pricing maintenance activities the cost estimator should take into consideration issues relating to downtime, disruption of services, non-availability of the service or asset, temporary decant costs and potential loss of income due to unavailability of use. The next step the analyst should decide on is how to price the identified maintenance and replacement schedules. Normally, in PFI/PPP schemes the client specifies in the FITN documentation how to present the estimated budget for maintenance and replacement. The actual prices for maintenance and replacements may be obtained from similar assets, specialist suppliers, contractors and subcontractors. In most cases a client specifies that all building components should be measured and priced for maintenance, operation and disposal. Some elements may prove to have a significant cost and risk attached to them. These ought to be concentrated on to make sure their allocated budget stays within the margin allowed by the client. Having said this, the author's preferred method for pricing maintenance at the design stage is to estimate these costs as a percentage of the capital cost of the component concerned. The replacement costs are straightforward to estimate. The analyst is only required to modify current costs for inflation. Based on these principles, the estimates in Table 9.8 are developed. Estimates shown in Table 9.8 are based on historical data from past similar projects. Data for replacement and decommissioning are hypothetical and used for illustration purposes. The data has modified to take into account all risk aspects and adjustment to inflation, location, market conditions, etc. The method of modification is well documented in other sources (Flanagan *et al.* 1977). Annual operational and maintenance costs are based on current costs of similar facilities. The frequency of alteration and replacement is also extracted from similar projects and they are mainly related to equipment and flooring changes.

To further illustrate the cost involved in maintenance, data from a sample of acute hospitals are presented in Table 9.8 and explained in the following sections.

Engineering maintenance

In developing and estimating the cost of engineering systems in hospitals, the analyst should take into consideration the total pay and non-pay cost for the provision of engineering maintenance services. This includes labour costs for directly employed and contract staff including contract support costs, fees, materials and pay elements for directors, senior managers and all associated staff in maintaining the fixed and portable engineering assets. Expenditure will also include costs relating to the employment of staff belonging to an external organisation (including PFI work). It includes the cost of all labour and materials incurred in maintaining items such as calorifiers, generators, security systems, fire alarm systems, building management systems, boiler

Table 9.8 Life cycle replacement

Capex	*18377697*				*Annual periods at 0% compound interest*									
Concession	*30 years*	*Design service life*	*Remaining life at handback*	*Replacement cycle*	*%*	*cost*	*1*	*2*	*5*	*10*	*15*	*20*	*25*	*30*
GFA	*5000 m²*				*Capex*	*£'000*	*£'000*	*£'000*	*£'000*	*£'000*	*£'000*	*£'000*	*£'000*	*£'000*
Services	Rising mains	30	0	replace @ 30	0.27	50								50
	Storage tanks	60	30	replace @ 60	0.11	20								
	Pumps	20	10	replace @ 20	1.09	200						200		
Plumbing and internal drainage	Pipework	30	10	replace @ 20	3.26	600						600		
	Sanitary ware	20	10	replace @ 20	1.63	300						300		
	WC pans	20	10	replace @ 20	0.82	150						150		
	Urinals	20	10	replace @ 20	0.44	80						80		
	Taps	20	10	replace @ 20	0.19	35						35		
	Valves	20	10	replace @ 20	0.30	55						55		
	Waste soil overflow and vent pipes	20	10	replace @ 20	0.82	150						150		
	Fuel tanks	60	30	replace @ 60	0.00									
	Boilers	25	20	replace @ 25	0.27	50							50	
	Burners	25	20	repaire @ 15	0.11	20					20			
Heating and ventilating	Air plant	25	20	repair @ 10	0.84	154				154				
	Pumps	20	10	replace @ 20	0.19	35					35			
	Water chillers	20	10	replace @ 10	0.44	80				80				
	Pressurised units	25	20	replace @ 10	0.33	60				60				
	Fan coil units	10	0	replace @ 10	0.07	13				13				
	Radiators	30	0	replace @ 30	0.65	120								120
	Ducts	60	30	replace @ 60	0.00									
	Valves	20	10	replace @ 20	0.54	100						100		

(Continued Overleaf)

Table 9.8 Continued

Capex	*18377697*				*Annual periods at 0% compound interest*									
Concession	*30 years*	*Design service life*	*Remaining life at handback*	*Replacement cycle*	*%*	*cost*	*1*	*2*	*5*	*10*	*15*	*20*	*25*	*30*
GFA	*5000 m²*				*Capex*	*£'000*	*£'000*	*£'000*	*£'000*	*£'000*	*£'000*	*£'000*	*£'000*	*£'000*
Electric power and lighting	Fittings	15	0	replace @ 15	0.95	175					175			
	Electrical switch panel	30	0	replace @ 30	0.44	80								80
	Wiring	60	30	replace @ 60	0.00									
	Fire alarm/bell system	15	0	replace @ 15	1.00	184					184			
	Back-up generator	15	0	replace @ 15	0.82	150					150			
	Medical gas	25	20	replace @ 25	1.76	324							324	
	Lift installations	25	20	replace @ 25	2.33	429							429	
External works	Drainage/water (UPVC)	20	10	replace @ 20	2.45	450						450		
	Roads and landscaping	20	10	repair @ 20	0.85	156						156		
External Wall	External structure	60	30	repalce @ 60										
	Walls	60	30	repointing @ 20	0.44	80	0	0	0	0	2	80	0	2
	Cladding	60	30	repair @ 30	0.28	52	0	0	0	3	4	5	5	52
	Flat roofs	20	10	change @ 20	2.04	375	0	0				375		
Roof	Pitched roofs	60	30	repair @ 20	0.33	60	0	0				60		
	Roof lights	60	30	repalce @ 60	0.01	2	0	0						
	Lean-to roofs	60	30	repair @ 20	0.11	20	0	0				20		

Windows and external doors	External windows	60	30	repair @ 20	0.14	25	0	0	5	5	4	25	2	5
	External doors	60	60	repair @ 15	0.08	15	0	0	1	4	15	2	1	3
	Curtain walls	60	30	repair @ 30	0.90	165	0	0	3	5	4	3		165
	Glazed screens	60	30	repair @ 20	0.16	30	0	0	2	3	4	30	0	4
Internal walls and doors	Internal partitions	60	30	repalce @ 60										
	Internal doors	60	30	replace @ 20	2.31	425						425		
	Stairs	60	30	repair @ 15	0.46	85					85			
	Balustrades	60	30	repair @ 15	0.05	10					10	15		
	Ironmongery	20	10	replace @ 20	0.08	15								
Internal finishes	Floor finishes	10	0	replace @ 10	6.46	1188				1188				
	Wall finishes	10	0	replace @ 10	1.25	230				230				
	Ceiling finishes	10	0	replace @ 20	3.23	594						594		
Decoration	External decorations	15	0	change @ 15	0.49	90					90			
	Internal decoration	10	0	change @ 10	0.97	178				178				
Fittings and fixtures		20	10	replace @ 20	4.63	850					850			

plant, electrical installations, and distribution services equipment within ducts, walkways and below ground plant areas, etc. Any roof-mounted equipment associated with internal equipment, e.g. split air conditioning units, is to be included in this category. Infrastructure associated with the engineering plant that runs external to the building/s should be included, e.g. pipework such as steam mains, distribution cabling (HV and LV) etc. Costs associated with medical equipment, pathology, microbiology diagnostic equipment, direct patient connected equipment, IT equipment and storage costs are normally excluded. The cost analysis of maintaining current engineering systems shows that value of this cost can fluctuate between £6 and £117 per OFA. This cost is significantly affected by the age and frequency of replacement of the engineering systems.

Building maintenance

The pricing of this element is related to total pay and non-pay cost for the provision of building maintenance services. It includes costs for directly employed and contract staff including contract support costs, fees, materials and pay elements for directors, senior managers and all associated staff in maintaining the buildings inclusive of inspections, surveys and work required for the preservation, repair and replacement with equivalent contemporary items of existing building components, plant services or equipment, preventative maintenance and follow-up remedial work and upkeep of external works. Other issues to be considered in estimating building maintenance include specialist inspections and the cost of materials, spares, and scaffolding. Historical data shown in Table 9.6 from current assets indicate that the cost of building maintenance for acute teaching hospitals could reach up to £50 per OFA or over. Obviously this rate will be dependent on several factors such the rate of inspection, age of the building, rate of monitoring and the rate of preventive maintenance.

Estimating disposal and decommissioning budget

There are several ways of dealing with this issue. At the end of the contract concession period the service provision could cease to exist and the asset could be sold, generating an income from the selling of the asset. The other option is the demolition of the asset which will result in estimating the cost of demolition and income from selling or recycling materials and components. However, one expects in the majority of cases, if not all, the asset will revert to the public sector at the end of concession period for a pre-determined sum. However, from value of money point of view one expects that the client will have already made arrangements in the project agreement to have the option to take the ownership of the asset at the end of the contract period without additional payment. If this is the case, then the provider cost analyst should not make allowance for the residual value in his computation assumptions. In some cases the provider and purchaser may choose to enter into a new services provision contract. In this situation the residual value of the facility might be determined from the difference between the net value of the existing facility and the net value of a hypothetical project with a new facility. The cost analyst should also bear in mind the cost associated with the disposal of the components being replaced during the contract concession period. It is important to point out here that in some circumstances the decommissioning of some of the infrastructure facilities could result in a huge liability for both provider and client, e.g., nuclear power plant.

Summary

This chapter has presented the process of cost planning at pre-financial close of PFI/PPP projects from the provider's point of view. Traditional cost planning processes are linear and exclude a large proportion of WLC budget planning aspects that are essential in PFI/PPP schemes. The challenge for the cost analyst is to develop whole life cycle cost plans with a minimum of detailed design and information. Cost plans are negotiated at the pre-financial close stage. At the FC stage, all prices, funding terms and payment mechanisms are confirmed and fixed.

Chapter 10

Cost variation control at pre-financial close and construction stages

Introduction

Construction control and monitoring require a systematic approach to inspections and compliance confirmation. Clear procedures should be outlined and put in place to carry out such an import task. The level of project control normally includes aspects of cost management, scheduling, and change control and risk management. Risk management is discussed in Chapters 12 and 13.

This chapter examines variation at pre-financial close and construction stages. First, the chapter examines the mechanism for setting control processes and benchmarks. Second, it describes the steps that are used in controlling costs. Third, the chapter introduces the process utilised in construction inspection of PFI/PPP schemes. Fourth, the chapter looks at time and scheduling performance, then the chapter explores design quality issues in PFI/PPP procurement. The chapter next discusses the issue of cost performance and control. The process of commissioning PFI/PPP facilities is introduced and explained. Finally, the chapter explains the processes and methods for dealing with change and variations during the construction stage.

Setting control benchmarks and mechanisms

The purpose of setting up control benchmarks is to assist PFI/PPP stakeholders to identify key performance indicators to which providers and clients might aspire. Setting control benchmarks ought to empower PFI partners to identify the key performance indicators that will assist all parties to maximise their performance and deliver value for money for their own organisations and others (see Chapter 3). Sohail *et al.* (2002) found that the reason for lack of performance monitoring is the non-availability of reliable performance indicators. Barber (2003) explains the general purpose of benchmarking as the process of investigating and learning from the best in a class to develop strategies for improving and changing the management of projects. According to Baker, benchmarking is one of most responsive evaluation tools for creating a learning project organisation that is receptive to both external and internal best project management practices. Hall *et al.* (2003) see the purpose of benchmarking as the process of comparing performance at the completion of the projects to a set of construction projects. The comparison is based on criteria such as cost and time predictability, degree of integration with the supply chain, value for money and sophistication of quality control systems.

In the PFI/PPP case, the benchmark could be the level of performance of the provider in delivering design, construction and operation of the scheme to the agreed service specification. The performance of the purchaser in terms of service specification, etc., should also be part of the benchmarking scheme. Li *et al.* (2001) stated that effective benchmarking should be founded as a partnership where contracting stakeholders can all benefit through sharing knowledge and improvement of their individual businesses. For the purpose of control in the PFI/PPP procurement context, a control benchmark can be defined as performance indicators for tracking the performance of the scheme throughout the stage of the procurement process and concession period. Sohail *et al.* (2002) state that lack of performance monitoring is due to the non-availability of reliable performance indicators. Hence, it is imperative to develop and implement control benchmarks that are able to give a clear account of the performance of PFI/PPP schemes.

Depending on the stage of the life cycle of the project development, a wide Variety of performance indicators can be employed for this purpose. The following indicators are of most relevance in the context of PFI/PPP projects:

At pre-financial close, the following are important:

- forecast of clients' outturn costs;
- bidding time;
- bidding cost;
- clarity of output specification;
- variation in output specification;
- value for money outturn;
- risk allocation;
- cost reliability/certainty;
- negotiation period.

At the construction stage the following are important:

- cost (i.e., target cost against actual cost comparison);
- clients' outturn costs;
- variations;
- level of compensation events;
- scheduling and programming performance;
- technical competency;
- quality for both design and construction, including defects;
- health and safety;

- dispute resolution;
- environmental issues;
- risk management.

Control mechanisms

The pre-requisite for efficient and effective control and monitoring of PFI/PPP schemes is the development and establishment of key control benchmarks or baseline performance indicators. The minimum requirement is the establishment of a clear baseline plan with clear performance indicators that should be achieved at every sign-off gateway stage or milestone of PFI/PPP schemes development. Figure 10.1 illustrates the process of an effective control mechanism. The process starts by defining the objectives and setting control benchmarks and desired performance. The performance indicators should be related to the scope and objectives of the scheme requirements, contractual aspects, schedules and budgets. A procedure should be set up and agreed by all stakeholders on how to collect data and interpret the measurement of the actual performance. The actual performance is then compared against agreed benchmarks. Any indication of variation from the specified baseline performance should be highlighted and analysis should be carried out to find the underlying causes and to develop appropriate strategies to deal with the variation. A programme of agreed action should be put in place to bring the project back to the agreed performance. For this process to succeed, stakeholders should initiate the control process as soon as possible. Regularity of monitoring and concentration on those aspects of the project that are critical to success of the project are also essential.

The role of project control is to provide the PFI/PPP stakeholders with a clear objective view about the status of the project progress from all aspects. If well

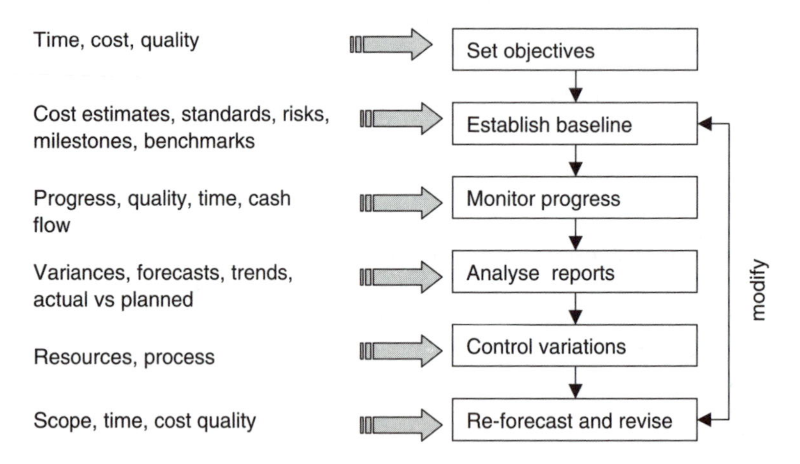

Figure 10.1 Control mechanisms.

developed and executed, such a system will add huge value to all parties by focusing the efforts of the management on critical issues to inform stakeholders about possible upcoming problems, so that they can devise strategies to deal with them. An effective control mechanism can be achieved through the following:

- periodical (time scale to be agreed between the contracting parties) contract progress reports. This report normally highlights issues relating to key performance indicators relating to productivity and quality of build (see Table 10.1).
- monthly progress reports by an independent body;
- periodical reports to financiers and insurers;
- periodical, cost and schedule reports;
- progress photographs at vantage points;
- progress and trend forecasts;
- exceptional reports on cost, time, quality and scope of critical elements of the contract that exceed or do not satisfy cash flow and quality profiles.

For example, in health PFI/PPP schemes, the design manual (NHS Estates 2002)

Table 10.1 An example of key performance indicators

Month May 2006		
Key indicator	*Method of measurement*	
1 construction programme	Target %	Actual %
Building 1 phase 1	60	55
Building 2 phase final phase	84	80
Non-works	90	90
Car park	30	25
2 Financial		
a client		
Issue of variations	no variation issued	
Building expenditure	on budget	
Equipment expenditure	on budget	
b provider		
Unitary payment	no payments made	
Unfunded variations to scheme	nil	
3 Health and safety		
Level of interference site services	nil	
Accident	10 minor incidents	
Enforcement by HSE reports	nil	
4 Independent tester assessment		
Quality of product/construction	satisfactory	
Adherence to specification	satisfactory	
Issues to be resolved	nil	

stipulates that progress control meetings should occur at monthly intervals throughout the construction period and should address the following issues:

- establish progress;
- record delays, delay events and payment events;
- identify access and interface problems;
- monitor the delivery and/or installation of equipment;
- respond to requests for extensions of time;
- record and report on the effect of variations;
- record the dates of valuations;
- record health and safety issues;
- record problems with workmanship;
- summarise any outstanding issues;
- review conflict between construction requirements/specification set by the client and provide proposed solutions.

Normally, according to the NHS Estates (2002), prior to the monthly meetings the provider is required to submit a detailed progress report summarising the following:

- a statement of progress of design and construction judged against the milestones set out in the agreed master programme of works;
- status of all consents and approvals;
- resumé of the reasons for any delay;
- actions to be taken to mitigate delays;
- variations requested by the provider;
- report on the impact of any variations requested by the client including any possible adjustment of the tariff payment – together with a schedule thereof;
- a progress report of the design development against the master programme identifying what action is being taken to mitigate any delays or accelerate the master programme;
- outstanding information required from the client and/or the provider company; health and safety issues; and a progress report indicating agreed baseline, actual progress to date and remaining durations for each activity;
- a schedule of warranties indicating progress in providing executed documents to the client.

Based on the analysis of data from the above comprehensive reports, clear procedures should be put in place to deal with any variation from the baseline forecasts. In this way quality of construction should be assured and a clear audit trail established of all aspects of the PFI/PPP deal.

Construction inspection

Most of the public authority manuals and guidelines on PFI/PPP place great emphasis on project management and control during the implementation of construction projects. One element of this is the mechanism established to ensure ongoing monitoring of project implementation. It is a requirement that the procuring authority must establish a project team to carry out the control and monitoring activities. Normally, this team is led by a director responsible for ensuring that the project execution plans are sound and compatible with value for money analysis and all management strategies set out in the full business case. From the client's point of view, it is the responsibility of this team to do the following:

- set out the plans, procedures and control processes for project implementation, monitoring and reporting;
- define the roles and responsibilities of all project participants;
- set time scales;
- identify the mechanisms for audit, review and feedback.

Monitoring of the programme, budget and quality of construction, including adherence to design, will generally be the responsibility of the providers and their contractors, though certification is carried out by an independent body. In PFI/PPP schemes, the client defines the overall project requirements. Based on these, the provider normally generates the project master programme, with key milestone dates and contract access, sectional completion and handover dates. In turn, the provider's contractor creates programmes to manage their contract scope, construction activities and subcontractors' work, which are consolidated and used to update the provider's master programmes.

One of the most important aspects of the construction process is the management of interfaces and site access arrangements for inspection. The procedure for this process is normally dealt with in the project agreement. This process usually necessitates clear roles and responsibilities to enable an effective liaison and co-ordination between the contracting parties to ensure rapid resolution of any access and interface issues and to ensure that construction works do not disrupt the day-to-day operation of the existing services. It is common practice that in PFI/PPP schemes the provider normally appoints a senior manager available on site to co-ordinate with the client who has the authority to receive and issue instructions in relation to those issues.

During the construction phase, the work is usually inspected regularly by the independent tester (IT) and progress observed by representatives from the client's organisation to ensure the building meets agreed contract technical requirements. The role of the IT is defined in the standard form of project agreement as to certify, on behalf of both parties, that the construction of the new facilities, as work progresses, has been completed to the agreed design and all other technical standards. In some cases the IT will also certify that the reconfiguration works and any other related works have been completed to agreed specification and standards. The appointment of the IT is in general a joint appointment and is paid for by the provider company. The IT will normally produce monthly assessment reports on the quality of construction and adherence to specification. Probably it will be a good practice if the contracting parties establish at the outset of the contract key quantitative assessment performance indicators (KPIs) to be used by the IT. An example is shown in Table 10.1. Information presented in this table will assist in closely monitoring all aspects of the technical aspects of the project. These KPIs are extremely important for controlling, monitoring and measuring the effectiveness and accuracy of the PFI/PPP deal, together with the value for money aspects. They are also useful in assessing the performance and efficiency of both the procurer and provider teams' performances. In this way quality of construction should be assured and a clear audit trail established

Schedule control

At the post-financial close the provisions relating to time control and delay events normally follow the standard form of project agreement. The project agreement places responsibility on the provider to complete the buildings on time and, until buildings are completed, he will not receive any payment. Time/cost overruns risk (except in the case of delay events and any changes approved by the client) in PFI/PPP schemes are normally transferred to the provider. Since the provider receives no payment until the services begin, there is a compelling incentive to ensure that buildings are completed on time. One also can argue that some of the standard project agreements do not provide compensation if the buildings are completed late. However, the contract normally includes a provision for long-stop date, and if construction is not completed by this time, the client can end the agreement and seek compensation.

It is the responsibility of the provider to make available to the client a detailed construction programme to enable the client to ensure that the construction is progressing as planned. In cases where the construction programme has fallen significantly behind schedule, the client can demand a revised construction programme and details of any strategies and actions to eliminate or reduce delay. In cases where the provider is unable to provide adequate quality control resulting in constructional defects, the client is entitled to monitor the construction process more closely, with the provider paying the client's additional costs incurred by this extra activity. However, if the monitoring process causes a delay to the provider's contractor's work through an unjustified inspection, this could constitute a delay event where the provider may receive compensation. But under the concession agreement the provider and their contractors are entitled to an extension of time or a revised completion date when they have been delayed by reasons for which they are not responsible – typically when design information is late, the employer instructs variations or there are weather problems. These are described as 'delay events', 'relief events' and '*force majeure* events', and are described in Chapter 5 and summarised briefly here.

Delay events and compensation events

The provider will be relieved only if a delay event has a material and adverse effect on the provider's ability to complete the construction of the building by the specified completion date. Delay events can only arise during the design and construct phase. In cases where the provider is entitled to relief as a result of a delay event, the compensation will be in the form of allowing the provider an extension of time, equal to the delay or impediment caused by the delay event, in respect of the completion date for completion of the works. According to HM Treasury (2004a); examples of these events are:

- an agreed variation to the works which prevents the works being completed on time;
- a breach of the client's obligations (in particular failing to give the contractor proper access to the site);
- interference from works carried out by the client which do not form part of the contract;
- access to the site where such access is not justified or reasonable for defective work monitoring;
- relief events such as an event of *force majeure* and relevant change of law.

In most PFI/PPP concession agreements, the compensation event regime is based on the clauses of a standard form project agreement and negotiations between the parties. Where a delay event is also determined as a compensation event, the provider is usually entitled to such compensation as would place it in no better or no worse position than if the relevant compensation event had not occurred (ibid.) The modality for compensation payment varies. But, in general, if the provider is able to provide finance for an amount equal to such compensation, the compensation is usually payable by the provider by means of varying the service payment. If the provider is unable to compensate for the delay, the client may pay the compensation by means of a payment of a capital sum. The standard project agreement states that where a *force majeure* event occurs before the actual completion date and the continuation of the *force majeure* event results in the termination of the project agreement, the client is required to compensate. Most of the concession contracts distinguish between compensation events, where the provider gets an extension of time and compensation; relief events, where he gets an extension of time only; and *force majeure* events, where both parties are relieved of their contractual obligations and, if the event persists, may be entitled to terminate altogether.

Relief events

Relief events are circumstances beyond the control of either party. Neither the client nor the provider can terminate the concession contract for failure by the other that is a result of a relief event. According to HM Treasury (2004a); the following are examples of relief events:

- fire, explosion, flood, riot and earthquake;
- failure by a utility provider or local authority to provide services;
- accidental damage to the works or the roads that service them;
- blockade or embargo;
- the discovery of fossils or antiques;
- unofficial strike or other industrial action affecting a large sector of the buildings maintenance or facilities management industry.

As with other delay events, the party affected by a relief event must take all reasonable steps to reduce the effects and continue to perform its obligations under the contract, otherwise relief will not be available (ibid.). It is common in most concession contracts that no compensation is payable in respect of such events, as they are not capable of being compensation events.

Providers in most cases are obliged to give notice and details of the delaying events, as failure to adhere to this stipulation could have a damaging effect on their right of recovery. Providers only receive compensation on matters that cause delay, as opposed to general disruption. This will expose providers to the high risk of not receiving reimbursement where the client causes it to have to re-programme or reschedule work or resources, without any delay being caused. This scenario will not arise in non-PFI/PPP standard contract and could lead to providers being in a worse-off position.

It is well known that in PFI/PPP schemes detail design development and construction run in parallel from financial close onwards. Figure 10.2 shows a typical design and build programme. Despite this, research shows that the construction of most PFI/PPP schemes is unproblematic and finishes on time. When the UK National Audit Office (2003a) examined 37 major PFI schemes, it found that 22 per cent were over budget in comparison to 73 per cent projects under conventional public procurement systems which were found over budget. The report found that 78 per cent of the projects surveyed were completed on the time specified on the contract. The investigation also found that 24 per cent of surveyed projects were delivered later than planned for at the financial close. The audit also showed that 8 per cent of the total surveyed projects hand overran by more than two months. But the report did not specify the extent of the overrun on time. In a previous report the audit office (NAO 2001) found that some 70 per cent of central government's construction projects were delivered late. Hence, the findings in this last report are considered as dramatic improvements in delivery on time cost performance. However, authors such as Sussex

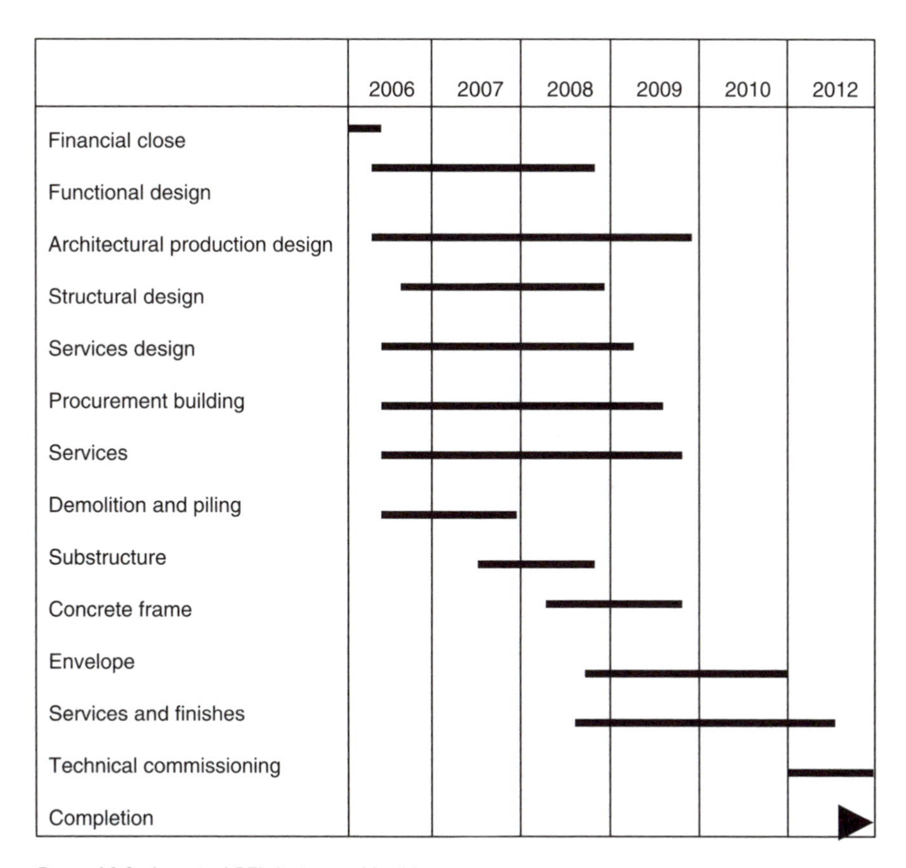

Figure 10.2 A typical PFI design and build programme.

(2003) and Edwards *et al.* (2004) argue that the measuring of time and cost overruns are based on false assumptions. Their argument is based on the fact that the large increase in cost and time of projects delivered under conventional public sector is due to the inequitable practice of comparing the outturn cost with the initial estimate made many years before the financial close. Their argument is that the outturn cost should be compared with the cost at financial close, taking into consideration all price inflation parameters. Sussex (2001) stated that if this approach has been adopted for value for money evaluation of PFI/PPP schemes, then these projects would be shown to overrun considerably on time and cost.

Quality control

Quality is an essential ingredient in building successful businesses and marketing. Organisations use performance indicators to improve the quality of buildings by adding more value through design, construction and operation of PFI schemes. Quality is the ability of each of the PFI/PPP procurement stakeholders' organisations to satisfy their client in terms of cost effectiveness and value for money services. If an organisation delivers quality to its customers and clients, in return, it can expect more profit, better relationships, more motivation, less waste, a competitive advantage and clients' repeated business.

Arditi and Gunaydin (1997) reported of an ASCE study of quality performance requirements in the construction industry. The study classified the requirements as follows:

- meeting the requirements of the client as to functional appropriateness and adequacy, completion on time and within budget, adequacy of life cycle cost and the provision of operation and maintenance strategies;
- meeting the requirements of the design professionals in terms of definition of scope of work, time allowance, timely decisions, etc.;
- meeting the requirements of the contractors in relation to performance of work on a reasonable time scale, timely decision, on change order, dispute resolution, etc.;
- meeting the requirements of regulatory authorities as to provision of health and safety and all other relevant regulations.

According to CABE (2005), design quality is a combination of functionality (how useful the facility is in achieving its purpose), impact (how well the facility creates a sense of place), and build quality (performance of the completed facility). Figure 10.3 shows the interaction between the design quality determinants.

CABE stated that every building design should confirm to the following design indicators:

- *Functionality*: the assessment of the quality and usefulness of the space. In PFI/PPP schemes, the client is responsible for the functional effectiveness of

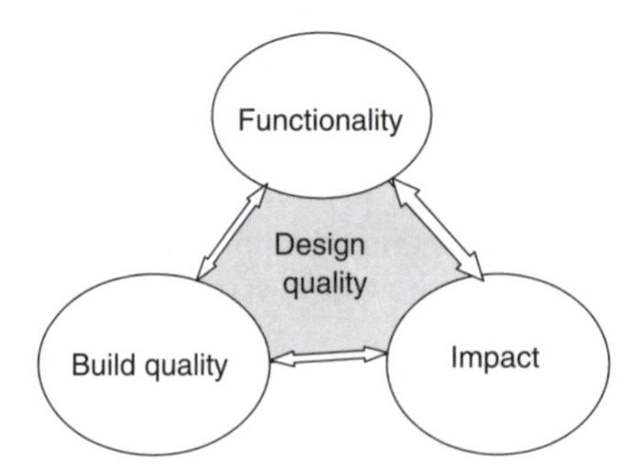

Figure 10.3 Design quality indicators.

Source: Adopted from CABE (2005).

the designs and, shortly before the project is signed, must review their service functionality and confirm to the provider at contract signature that the designs satisfy the client's requirements, whereas the provider is responsible for producing designs to meet the client's requirements as set out in the tender documentation. Hence, the risk of failing to translate the requirements of the client into the design lies with the provider.

- *Build quality*: the assessment of the performance of the building in engineering terms; its structural stability and the integration and robustness of the systems, finishes and fittings. In PFI/PPP schemes any defect in build quality resulting from construction that does not satisfy the client's construction requirements (i.e., the performance and functionality of the buildings ought to remain at the agreed standard throughout the life cycle of the concession agreement) is the responsibility of the provider. The provider will bear the cost of any changes or deviation from the agreed standards.
- *Impact*: the assessment of the effect the building has on the local community and environment, and the ability of the building to create a sense of place. The project agreement places a duty on the provider to use reasonable skill and care in its design and, in particular, to ensure that the building meets the client's operational requirements in terms of energy efficiency, etc. If, after monitoring this for a specified number of years, poor design means these standards are missed, then the client may be entitled to compensation.

The benefits that businesses should derive from a properly implemented quality system are:

- improving customer satisfaction;
- improving efficiency;
- improving effectiveness;
- reducing rework and waste;
- creating a well-planned business;
- adding credibility to stakeholders' organisation;
- enabling the stakeholder to compete on an equal basis with others;
- increase in value, both tangible and intangible;
- attainment of strategic and project objectives;
- increased learning;
- client satisfaction;
- beneficial social and environmental impact.

Quality failure could be attributed to the following factors:

- incomplete application;
- poorly communicated brief;

- incomplete or inaccurate design information;
- late design information;
- poor programme;
- accumulated delays;
- poor communication between the procuring authority and the bidders;
- unclear objectives;
- lack of real sense of shared purpose;
- lack of realistic expectations;
- failure to set up clear reporting and communication structures;
- inappropriate appointment of consultants to the wrong brief or at the wrong time;
- adoption of an inappropriate procurement form;
- failure to develop a comprehensive brief;
- failure to ensure scheme designs and design development meet the brief;
- poor workmanship;
- failure to monitor and manage the construction process.

Inadequate quality control could result in the following consequences:

- planning delay;
- redesign;
- low design productivity;
- design queries;
- out of sequence working;
- low operative productivity;
- rework;
- acceleration of design costs;
- poor service delivery;
- higher unavailability rate;
- lower return from the operation of PFI schemes;
- higher operation costs;
- higher life cycle costs.

One can argue that the quality of design is directly related to capital spent on building elements at the detailed design development. It is suggested (Pringle and Cole 2004) that there is a tendency for design and build contractors to try to minimise the capital costs by adopting lower-quality specifications for finishes and services as well as using strategies to reduce areas or volumes and omit some of the design features. If we accept this argument, then it is possible in some PFI/PPP schemes that design and build contractors may try to minimise the capital cost of the project to maximise the profit from the construction operation. This could have far-reaching consequences during the operation stage. Probably the operating company will have little recourse against the contracting company if the client's requirements are met, but the operation aspects of the facility are compromised. This scenario could arise from conflicts if the operating company is different from the design and build contractor. The current thinking is to keep the solution simple (basically using tried and tested methods) in order to make commissioning easy, to minimise the risk of payment deduction and to make maintenance cost efficient. Obviously this practice is contrary to the whole life cycle design and innovation paradigms. All stakeholders should aim to pursue whole life operational strategies that assist in minimising all aspects of operational expenses. The problem here is that the selection of most of the equipment and building element, in most cases, is not subject to any rigorous value engineering analysis prior to financial close.

It is well recognised that design quality has not been sufficiently emphasised under PFI/PPP procurement systems, and the public sector bears all or most design risk. According to Pringle and Cole (2004):

> It is a widely held view that design quality, both in its aesthetic and its more holistic sense, has generally not received sufficient emphasis, often being relegated to a minor consideration compared with contractual, commercial and deal-closing priorities.

A detailed survey of 27 UK contracts by Pepperell (2002) revealed a number of elements that lead to increased construction costs, delays and cost claims, suggesting that some designers did not understand or know the correct way to present their design material, and the consequences that might follow. The main highlights of these findings are:

- 40 per cent of all drawings contained either a design error, a drawing mistake, or required revising;
- 23 per cent of all drawings had missing information, or there was no specification to support the designer's intention nor define the expected performance purpose minimum requirement;
- 19 per cent of all documents had differing or conflicting information, giving rise to errors in interpretation;
- 10 per cent of all drawings and documents contained obscure or confusing information (a clear mismatch in effective presentation and basic understanding);
- 6 per cent of all drawings, designs or details were not understood in regard to intention or purpose;
- 2 per cent of all drawings that had been revised were still in a form that could not be understood by those to whom they were issued.

The Audit Commission report (NAO 2003c) also found that the quality of schools built under traditional public sector procurement was better than PFI schools. A report by BRE (2002) for the Audit Commission (NAO 2003c) on the design quality of PFI schools has concluded the followings:

- All schools investigated need significant room for design improvements, particularly at a detailed level.
- PFI schools were generally of a lower design quality than traditionally procured ones.
- The PFI schools scored significantly less in the areas of architectural design quality, cost of ownership, and detail design.
- In the areas of building service, design quality and user productivity, indicators showed little difference between PFI and traditionally procured schools.

The RIBA (2005) is proposing a smart PFI model designed to deliver good standards of design and quick competitive bidding. RIBA claims that their proposed model will improve design quality while simultaneously reducing the cost and time required to complete the PFI bidding process. The main features of the RIBA's exemplar model lies in the concept that the public sector takes the sole responsibility for the development of need, strategic planning, quality objectives, concept design and determination of affordability. This could well present a step in the right direction in solving the problem of design quality in PFI/PPP schemes. However, the RIBA proposed exemplar model is lacking in many aspects, especially in the integration of the issue design, construction, operation and commercial interests at all stages of the PFI/PPP procurement process. All these essential elements of the PFI/PPP deal are crowded into a short period of the PFI/PPP bidding and construction process. Design is just one part of the PFI/PPP process that is also competing for time and funding against the decisions being made on construction, operation and commercial interests. Maybe at end of the day, the project sponsor may opt to select a solution that could minimise the FM cost, but this could be the worst design yet this design satisfies the client output specification. Probably, it will be very hard to achieve the intended quality results and better value for money, without a robust integration of all PFI/PPP processes.

Controlling costs

Cost control at both pre- and post-financial close should be an integral part of the agenda of value for money delivery. An effective cost process will enable both the provider and the client to do the following:

- set a firm cost plan;
- produce strategies to achieve acceptable design quality;
- monitor the effect on the capital and whole life cost of building elements in both pre- and post-financial close phases.

A similar system should also be put in place for the financial management of facilities and delivery of services at the operation stage. However, one should recognise the fact that at the post-financial close the financial control of PFI/PPP schemes is an inseparable part of the project agreement.

According to Betts (1992), an effective financial control process should have the following characteristics:

- the ability to integrate the requirements for time, cost and quality by allowing informed trade-off decisions throughout the life cycle of scheme development;
- a dynamic rather than administratively reactive process in the attainment of stated performance;
- the ability to develop and implement changes at the earliest stages and throughout the life cycle of the project.

The main thrust of Betts' first point might be difficult to implement in PFI/PPP procurement process. The three performance indicators, time, cost and quality, are normally put together at the pre-financial close, when the detailed design development is incomplete. This problem has led other authors such as Atkinson (1999) to suggest new indicators by combining time, cost and quality into a single performance measure in conjunction to other measures such value (benefits) to stakeholders. Other researches (Gardiner and Stewart 2000) have advocated the use of NPV as a control mechanism. Their rationale is that at the implementation stage managers should take into account the time effect on cash flow in case of project delays. Their arguments are that in absorbing the occurred delay that managers will benefit from analysing the impact each alternative solution will have on the project's NPV. The author supports the use of NPV techniques as a cost control mechanism in PFI/PPP schemes. This could be achieved through continuous review of the financial case of the scheme at post-financial close to authenticate the impact of changes and variations on the NPV of scheme. If this process is implemented, without doubt it will lead to improving our knowledge and understanding of value for money creation throughout the life cycle of PFI/PPP schemes.

The UK Audit Office investigation into the construction performance of PFI projects found that all the surveyed construction companies had experienced cost overruns on some their PFI projects. The reported cost overrun is between 10 and 25 per cent. According to NAO (2003a), the reasons cited for these substantial construction increases include:

- weather conditions;
- unforeseen ground conditions;
- labour problems;
- changing building regulations.

According to the same report, these huge cost increases are generally absorbed by construction companies. This fact is portrayed as evidence of risk transfer working in the best interest of the client. Normally, the client transfers the construction risk to the project company who in turn transfers the construction risk to design and building construction subcontractors.

Commissioning

In most PFI/PPP schemes the commissioning process and responsibilities are defined by the project agreement. For example, the UK standard PFI contract stipulates the following:

- The procuring authority must provide the project company with a commissioning programme identifying all essential issues before and after completion.
- Prior to completion, the provider must notify the purchaser of the date on which completion will occur.
- The IT will inspect the building and notify the contracting parties of any outstanding issues.
- When the outstanding issues are resolved, the IT can issue a certificate stating the building is complete (the IT normally issues a snagging list, but this should not stop the IT from the declaring the building is complete).

It is common practice for the client normally to form a commissioning team, comprising different managerial and operation skills and users of the new facilities, to be responsible for bringing the new facilities smoothly into operation. Usually the commissioning management and programme arrangements are developed by the client well ahead of the building completion. The arrangements will include:

- preparation of a commissioning and services planning, including risk management and operation strategies;
- preparation of handbooks and operational guides for operating the new building;
- detailed equipment installation schedules;
- training and induction for new facilities;
- set-up of service PFI monitoring team;
- set-up of evaluation procedures and strategies.

Dealing with cost scope, service and cost variations

A report by the UK National Audit Office (2003a) found that there had been a price increase in the PFI schemes investigated. The report found that most of the changes were instigated by the client. The report confirmed that the price changes mainly related to further work, which had not been part of the original specification, or additional or improved facilities or changes to the function of a building.

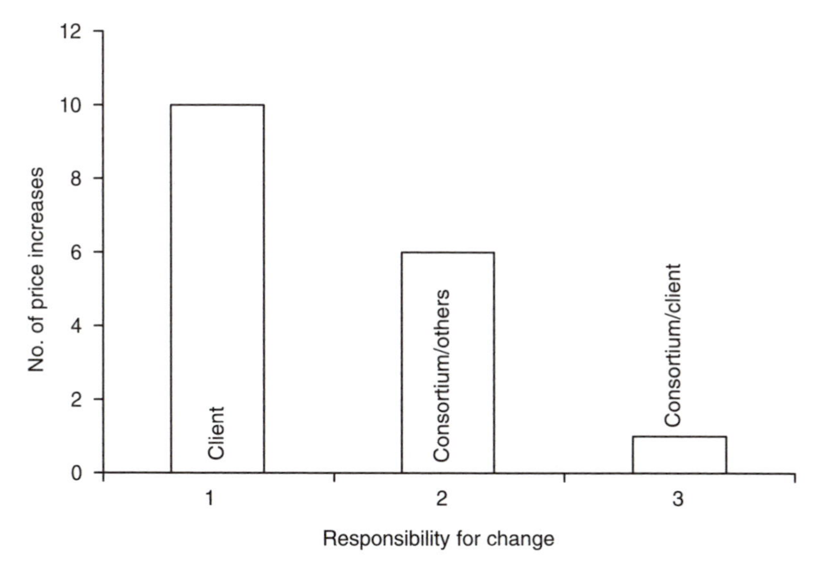

Figure 10.4 Responsibility for price increases.

These findings are shown in Figure 10.4 and Figure 10.5. The figures show the sources and reasons for price variations at post-financial close. The data confirm that most of the variations in construction cost increases are due to changes initiated by the client. Reasons cited for price increases include (NAO 2003b):

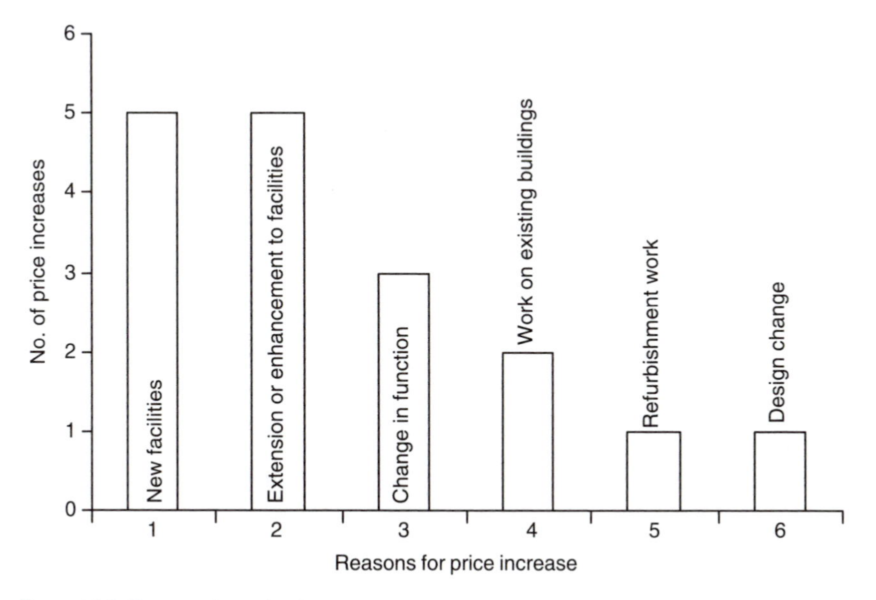

Figure 10.5 Reasons for price increase.

- new facilities;
- extensions or enhancements to facilities;
- change in function;
- work on buildings which had not been expected to be retained;
- refurbishment work;
- design change.

To allow for future flexibility, as circumstances and needs change, the potential to negotiate variations to a PFI contract is essential. The rule of thumb is that, it is best for the client not to issue any variation orders after financial close. However, this is somewhat unrealistic; one should anticipate variations across the life cycle of the project concession. That it is why the UK Treasury has reported that one in every four PFI contracts has been adjusted during the construction period to take account of changes in scope required by the customer. An amount of flexibility should be included in PFI/PPP contracts to allow for any variations or additional work. Project agreement clauses covering variations must allow for a cost-efficient and rapid processing of both high-volume, small-value variations, less frequent and more major changes during the life cycle of the project.

One of the main potential drivers for variation is failure in service and design specification. If variations are due to failure in specification, then the client has the choice to finance the variations or adopt different strategies. A commonly used strategy is to downgrade the specifications in value engineering terms without affecting the user or with a minimum effect on the operational aspect of the facility. Another option is for the project company to increase the value of the design and build contract and decrease the value of the operation contract. The implementation of the client variations could lead to some of the contract details being re-negotiated. For example, the new variations might have an effect on increasing the need for maintenance and life cycle replacements in the concession period. It is important to point out here that any renegotiation of the contract detail may lead to compromising the completion date.

Variations in design changes can be clustered in the following three categories:

- Design is correct in all aspects, but the client desires to change the specification.
- Design is accurate, but it needs fine tuning and minor changes.
- Design in incorrect and re-designing is required.

Depending on the category of change required, the design variation can be classified as variations or design developments. Design variation is a design change that departs from the agreed specification. Design development variation is mainly related to change in the 1:50 drawings that are within the above clusters of design change. However, since most of the PFI/PPP schemes are procured on a design and build basis, one should expect that the contractor will take the responsibility for redesign arising from design developments and accordingly they have already made allowances in their fixed-price design and build contract.

All variations are normally dealt with and priced in accordance with the relevant clauses in the project agreement. To minimise variation throughout the life cycle of PFI/PPP schemes, the UK HM Treasury in *Standardisation of PFI Contracts*, Version 3 (2004a) has recommended the following strategies:

- Low price contracts may give departments little flexibility as the consortium may have little scope within the terms of the contract to absorb unforeseen changes.
- To preserve flexibility, departments should think carefully about mechanisms which will maintain value for money. These include agreed prices for defined options for additional work, agreed rates or profit margins for unspecified further work, the rights to benchmark the pricing of additional work and open book accounting.
- Changes during the construction or development phase should be kept to a minimum unless a long period of time is scheduled to elapse before the service commences.
- The procuring authority should serve a notice setting out the intended change and should require the project company to provide an estimate of the technical, financial, contractual and timetable implications of the change within a specified period.
- The project's financiers, in particular, may not allow the project company to agree to any change which would increase project risk, financing risk or reduce the rate of return.
- If additional works or variation is needed, extra finance may be required. The most important aspect here is that the client should ensure that the finance provided for variations represents good value for money. The finance may be provided by the client alone, the project company alone or in combination. The client has the final say on the funding arrangements and the party responsible for obtaining finance must use best efforts to obtain it. Additional work and variation can go ahead if finance is available and represents value for money.
- The procuring authority should ensure value for money by paying a reasonable price to implement the change by including provisions within the project agreement as follows:
    - the project company has the duty to mitigate the costs;
    - transparency of information on costing;
    - If the change is to be implemented by a sub-contractor, competitive quotes should obtained and, if possible, competitive tendering should be sought in respect of any capital works required to implement the relevant change in service.

    - If the change is to be implemented by the project company, that the cost is benchmarked against prevailing market rates.
- Any increase in operating costs resulting from client change, during the construction or operation phases is normally met by an increase in the unitary charge. However, if changes result in reduction to the provider's costs (capital or operating costs), then an appropriate reduction should be made to the unitary charge.

It is common knowledge that any price variation after the FITN stage may occur without any market competition. This could have a significant impact on the value for money analysis carried out by the purchaser and probably in most cases could result in variations that are not value for money. This view is supported by findings from a report on managing PFI relationships by the UK National Audit office (NAO 2001). The report stated: 'Some public sector project teams which had experienced a decline in value for money after contract letting stated that the decline was due to high changes for additional services.'

The variation process

The variation process normally depends on the clauses included in the project agreement and can vary from project to project depending on the complexity of the project and financial agreement. However, the following are broad guidelines of the process that is stipulated in the standard PFI contract.

- Identification of reason of variations and parties that want to make the variation.
- Initiation of the variation process through variation enquiry forms.
- Detail of the variations, i.e., whether it is related to works or service.
- If the variations are issued by the client, the project company is obliged then to provide initial indication of cost, timescale and the availability of finance for change.
- If the variations are issued by the project company, the client has complete discretion to agree or refuse the alteration. The contractor must also demonstrate how it will pay any extra cost of the proposed variations and that it has the appropriate consents from the consortium to proceed with the variations.
- The project company must inform the client if it objects or agrees to the proposed variations. The provider can object to variations on the grounds of health and safety concerns, variations that would result in a breach of law, and if the variation would fundamentally change some key aspects of the project agreement.
- If the provider agrees to the change, it must provide a detailed estimate of costs and timescale of the proposed variations.
- The provider also confirms whether the proposed change would lead to the

alteration to the payment mechanism and any relevant terms of the project agreement.

- The contracting parties, including the financiers, must reach an agreement on cost, timescale and contractual issues.
- The client will take the final decision, based on value for money paradigm, on whether to go ahead with the proposed variation.

Summary

This chapter has examined cost variation at the pre-financial close and construction stages. The author advocates the use of control benchmarks to empower PFI/PPP partners to maximise their performance and deliver value for their own organisations and to others. The author proposes the use of benchmarking to track the performance of PFI/PPP schemes throughout the life cycle of the project. The role of cost control of PFI/PPP projects is to provide the contracting parties with a clear objective view about the status of the project progress from all aspects. Construction inspection is carried out to assure the quality of construction and provide an audit trail of performances. Provisions relating to time control and delay events follow the standard form of project agreement clauses. This chapter has shown that design quality has not been sufficiently addressed in PFI/PPP projects. Continuous review of the financial case of PFI/PPP schemes at the pre- and post-financial close to authenticate the impact of changes and variations on the NPV of scheme is indispensable for cost control. Most construction cost variations, in past PFI/PPP projects, are due to changes initiated by the client. All types of variations are managed and priced according to project agreements clauses.

Chapter 11

Monitoring and responding to cost variations at the operation stage

Introduction

Operational and maintenance costs play an important role in the procurement of PFI/PPP facilities. Construction stakeholders have become increasingly aware of the importance of early stage advice on the cost implications of design decisions. Decisions made during design can have a significant impact upon the future running and maintenance costs of buildings; this is especially true of high occupancy buildings such as public buildings. This chapter will introduce the mechanisms used in the management of operation and maintenance costs. First, it addresses the issue of operational cost benchmarking and explains the processes put in place to assess the technical and financial performance of PFI/PPP schemes. Second, it identifies the incentives to use condition surveys for maintenance management. Third, the chapter explores the stipulated procedures for controlling maintenance and repair costs. Finally, it examines the contractual systems established to control operational costs.

Setting cost control benchmarks for operation and maintenance

Measuring best value and performance has long been the intention of policy-makers, and central to this school of thought has been the application of benchmarking techniques. Benchmarking has been identified as a powerful vehicle for quality improvement and a paradigm for effectively managing the transformation of public-sector organisations into public-sector organisations of quality (Weller 1996). Benchmarking and performance measurement techniques have assumed a central role in public service delivery administration, no more so than in the UK. Specifically, UK government agencies have strongly advocated the use of key performance indicators (KPIs) as a tool for productivity improvement and modernisation. An essential part of the PFI/PPP paradigm is the requirement imposed on the client to monitor performance, user experience, risk allocation and pricing, and all other value-for-money drivers.

The aim of this exercise is to compare outturns with service expectations, and the results and lessons learned from the monitoring process may be used to improve future decisions on capital investment and the operational efficiency of public services. Hence, the purpose of setting control benchmarks should be towards identifying opportunities for improvement. This section will address the subject of setting control benchmarks and introduces the options available for benchmarking.

The process of benchmarking

There is a perception that if key performance indicators are put in place, managers will automatically follow them and everything will be fine. Unfortunately this is not the case in real life and benchmarking is an expensive dynamic process and intrinsically involves all PFI/PPP stakeholders who provide the information for the process and receive comparisons results. As shown in Figure 11.1, the process involves defining objectives, inputs, processes, analysis, outputs, feedback, monitoring and governance. The first step in the benchmarking process is the definition of the objective and scope accompanied by a statement of the anticipated benefits. This process is iterative, it is normally discussed at pre-financial close and it should be an integral part of the facility management agreement. The inputs are related to the identification and selection of the key

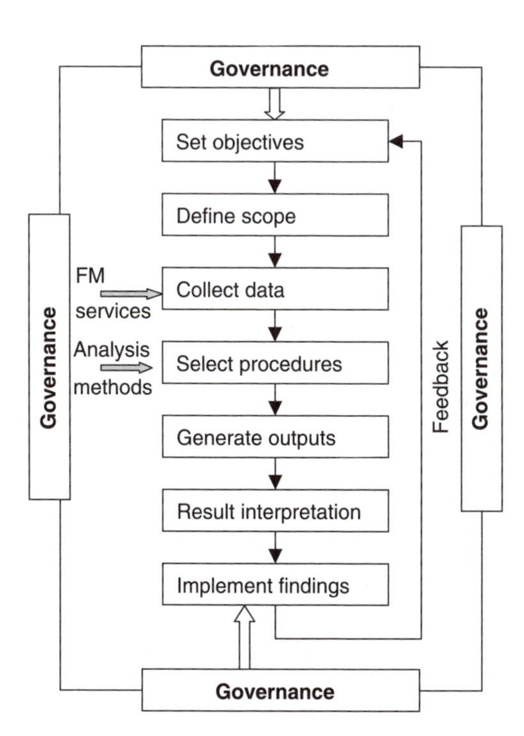

Figure 11.1 The process of benchmarking.

performance indicators that are going to be used to measure the performance of the contractor and the value-for-money benefits to the client. The key performance indicators are normally regularly reviewed and adjusted to take into consideration any service changes. In PFI/PPP health schemes, key performance indicators are normally set up for the following services (NAO 2003a):

- estates and maintenance, grounds and garden;
- portering, transport and internal security;
- switchboard and telecommunications;
- external security;
- linen and laundry;
- catering;
- domestic window cleaning and pest control;
- car parking.

The processes stage deals with the mechanisms to be used in planning and implementing the benchmarking process. Issues such as timing and frequency of data collection and procedures used to gather the required data are dealt with here. The analysis stage is concerned with the selection of the appropriate method and modelling techniques for analysing the collected data. To do this, there are a number of methods available, ranging from very sophisticated statistical methods to simple scoring systems. The choice of a method depends largely on the complexity of system or process to be benchmarked, although the author prefers statistical methods for the simple reason that statistical prediction can be used to analyse heterogeneous FM processes. The generation of reports/outputs phase deals with interpretation and communication of the results to all stakeholders. The analyst should look at measures of effectiveness and reductions in cost. It is important here that managers should act on the results to improve the efficiency of the FM processes. Feedback should be an integral part of the benchmarking process. The feedback is related to both informing the decision-makers and managers about the results and the efficiency measures that need to be taken and informing the process of setting targets, benchmarks and objectives. The monitoring process deals with the continuous evaluation of the benchmarking process and assessment of the efficiency of implemented decisions, including reporting structures and payback. Feedback from the benchmarking process provides an opportunity for facilities managers to learn. Through learning it is possible for FM stakeholders to become more effective and efficient in delivering FM services. Finally, the overall benchmark marking process must be managed through an effective governance structure. The process needs to be led by a strongly committed leadership, responsible for strategy setting process and policies as well as overseeing the whole benchmark process.

The biggest problem with setting cost control benchmarks for operation and maintenance purposes is the lack of a benchmark. In the absence of an industry-wide benchmark (or standard), service providers may aim for an average performance as defined in the project agreement. A great deal of research is necessary to gain an understanding of the key performance indicators relevant to a particular public purchaser. The large capital investment in PFI/PPP projects has led to increased interest in financial and non-financial performance indicators and efficiency drivers. One should not be surprised to find obvious differences in the performance indicators for schemes for health, education, prisons, defence, and transport.

Edward *et al.* (2004) listed the available standards for measuring and evaluating the operational performance of PFI/PPP schemes as (1) benchmarking against the original project agreement; (2) benchmarking against the public sector comparator; (3) benchmarking against conventional procurement methods; and (4) benchmarking against private sector performance. These are discussed in some detail in the following subsections.

Benchmarking against the original project agreement

It is anticipated here that the actual performance standards will be compared to the original output specifications. In order to have a robust benchmarking system, the output specifications have to be clearly defined and developed at the contract stage. The problem with this method of benchmarking is that the length of the contract over 30 years requires contracting parties to put in place robust procedures that will accommodate changes in requirements but allow for the continual provision of services during the contract period.

Benchmarking against the public sector comparator

It is a requirement that every procuring authority should prepare a PSC as a pre-condition for evaluation of the options at outline businesses case stage. The idea here is to compare the cost of the scheme using initial public procurement cost with the PFI/PPP option. The PSC is used as a reference case for evaluating value-for-money aspects of the proposed development. The problem with using PSC for benchmarking the operational aspects of PFI/PPP schemes is that the majority of PSCs are badly developed and significant errors were found in the assumptions and modelling in some of the PSCs. Also, most of the project parameters change during the financial close and operational phase of the deal. This will require an overall updating of PSC assumptions and computation to fit the new changes and contract conditions. This is not common practice at the moment. Such problems have led authors such as Edwards *et al.* (2004) to suggest that it is unlikely that the PSC will provide a good benchmark. However, one can argue that the operational aspect of PFI/PPP schemes can be evaluated against similar public sector projects using the actual yearly operating performance data to construct and develop real time benchmarks as shown in Figure 11.1.

Benchmarking against conventional procurement methods

The process of benchmarking PFI/PPP service delivery against conventional public sector service delivery methods involves comparing the service delivery in two completely different environments. The PFI/PPP sector has new facilities and different contractual staffing arrangements and working patterns, whereas the public sector facilities are mainly old and require a huge capital injection to bring them up to a reasonable standard. Also, staffing and working arrangements in the public sector are different, especially in relation to pay and pensions issues. Without a doubt this diversity in practice will have a significant impact on any performance measure that attempts to compare the two sectors.

Benchmarking against private sector performance

Some of the existing PFI/PPP sectors use this performance measurement scheme. The process is based on using operational costs and performance levels from private sector competitors of the service provider (Grimshaw *et al.* 2002). In theory this process seems logical and market oriented. However, the reality is quite different: first, it is very difficult to obtain cost and commercially sensitive information from competitors, second, most service contracts are not bid for competitively on the open market (the process is referred to as market testing) which makes it difficult to provide adequate comparison for benchmarking. Northcott and Llewellyn (2003) found that there are several problems in using the reference costs database for benchmarking. The problems include:

- lack of a standard against which reference costs can be compared;
- non-compatibility of many hospitals featuring in the index;
- lack of standardisation in costing practices.

Hence, the inability to examine current market prices makes it difficult to make a meaningful comparison between the PFI/PPP service providers' performance and their competitors.

Perhaps the way forward is to allow public sector organisations to benchmark between contractors in the same sector over time. This is important because some of the service provisions are radically different from others. For example, prisons are a unique case; it is not possible to compare them with schools or hospitals. Hence the author proposes benchmarking between contractors in the same sector, which will provide a platform for comparing like-for-like performances.

Benchmarking between contractors in the same sector

Measurement of the performance of PFI/PPP schemes is complicated by the fact that there are many outcomes of a different nature, some of which are difficult to measure; this problem is further affected by the fact that solutions to the performance measurement problem may require a multiple output which makes it difficult to identify the key performance indicators. It was also suggested that traditional measures of operational efficiency, such as shareholder value creation and profit maximisation are inappropriate in assessing the efficiency of non-profit-making organisations, such as the public sector (Chen *et al.* 2003).

Based on the above stated problems in setting control benchmarks, the author proposes the use of a cluster of services according to their function, i.e., schools, hospitals prisons, etc., to assess their performance. It is also possible to sub-cluster the PFI/PPP services by size, location or any other relevant factors. This paradigm is based on the idea that the best approach for evaluating operational costs is to study comparable facilities. The aim here is to develop a more comprehensive approach to linking the FM costs to an agreed performance in terms of service provision. These steps are necessary since the size and function of a facility can have a significant effect on the outcome of operating costs and services provided (Kirkham *et al.* 2002). If costs that underpin the inefficiency are not compiled within a comparable framework, the inefficiency variability that results may merely reflect different costing practices rather than varying levels of cost efficiency.

The underlying theme is to evaluate how PFI/PPP schemes can effectively convert infrastructure resources into different types of service provision. Hence, the aim here is to measure how efficiently the facility management costs (resources) are utilised in the production processes by the contractor to deliver value for money to the purchaser. The author proposes the use of Data Envelopment Analysis (DEA) methodology for performance measurement of PFI/PPP schemes. DEA is an operational research tool that allows the measurement of efficiency indices by solving linear programming problems. DEA has proved to be a useful tool for measuring the efficiency of organisations that have multiple input and output structures, such as hospitals. DEA has been used to analyse the efficiency of Air Force maintenance units, bank branches, primary, secondary and tertiary education, hospitals, fast food chains, post offices, etc. (Boussofiane *et al.* 1997). Since the initial development of the theory, a significant number of publications exploring both theory and application in public and private sectors have been published. Boussofiane *et al.* (1997) uses DEA to assess the level of technical efficiency of organisations privatised in the UK in the 1980s, with the aim of establishing if privatisation had an effect on the organisational efficiency. Puig-Junoy (2000) uses the method to partition input cost efficiency into its allocative and technical components in Spanish hospitals. Tsai and Molinero (2002) developed a variable returns to scale DEA model for the joint determination of efficiency within UK health services. A detailed discussion on how this method can be used to benchmark FM costs in hospitals can be found in Boussabaine and Kirkham (2006).

Project performance

Normally the project agreement sets out the process and principles by which technical and financial performance measurement systems would operate to measure the delivery of facilities management services. It is important here to point out that the core service, the delivery of education, clinical services, etc. to the public, is not part of the PFI/PPP contract. These issues are described and discussed in the following subsections.

Financial performance

Unlike conventional procurement of construction projects, in PFI/PPP projects the lenders have huge control on how the cash flow of the project is used and distributed during both the construction and operation periods of the concession. Lenders and investors protect their long-term investment in respect of the construction and operation of a private finance contract by setting specific financial tests and limits for the private contractors responsible for delivering the project (Audit Scotland 2002). According to Yescombe (2002): the cost control by lenders includes the following:

- setting out the order of priorities for use of income from the project (see Chapters 4 and 6);
- lenders make the project company set up separate operational reserve funds as security against any short-term cash flow problems and whole life cycle expenditure;
- control on distribution of dividend to investors is set on the basis that all operating costs, lenders' repayment and reserve accounts are paid first;
- dividend distribution to investors is made once all these expenditures are paid for;
- in some circumstances, lenders may allow investors to distribute dividends, provided a clawback is undertaken by the project sponsors if the project company develops cash flow problems.

Lenders will also require the SPV to demonstrate that it is able to meet its debt obligations in each and every year of the project. It does this through the use of financial ratios. In a classic example, when a principal subcontractor who was responsible for the design, construction and management of a PFI scheme experienced cash flow problems emanating from the failure of the principal subcontractor to secure due payments from a series of other construction projects, the lenders took swift action to take over and manage the principal subcontractor (4ps 2004a).

The responsibly for controlling the life cycle funds, used for the major maintenance and replacement of building components that are going to require significant expenditure over the contract period, vary from project to project.

In some cases, the project operator is responsible for managing this fund, whereas in other projects the SPV is responsible for managing and controlling this fund. In some circumstances the operator is responsible for procuring and managing the minor maintenance of components below a specified value as part of its FM service contract. Above this benchmark value, the SPV will procure the major maintenance and replacement, including its management. If the operator wants to carry out the work, it will have to bid for it in competition with other building maintenance contractors. Since there is a direct relationship between capital expenditure and operational costs, the stakeholder who is responsible for managing this relationship should have overall control of this fund.

The client has control over operating cash flow through payment deductions for underperformance. All project agreements state that the client is entitled to make payment deductions for unsatisfactory service performance (see Chapter 5). Figure 11.2 shows a real example of how a project agreement provided for the amount of the deduction to increase in line with significance of service failure. Research (KPMG 2005) has demonstrated that 61 per cent of private contractors had been subjected to deductions for either non-performance or unavailability. The size of an underperformance deduction varies and it is triggered if the level of underperformance exceeds a pre-specified threshold. The payment deduction could be made up to the full amount of the project company's normal charges for any particular service in any month (NAO 2002).

Market testing

The concept of benchmarking FM services is discussed above. Market testing is also used as a vehicle to ensure that value for money principles are adhered to

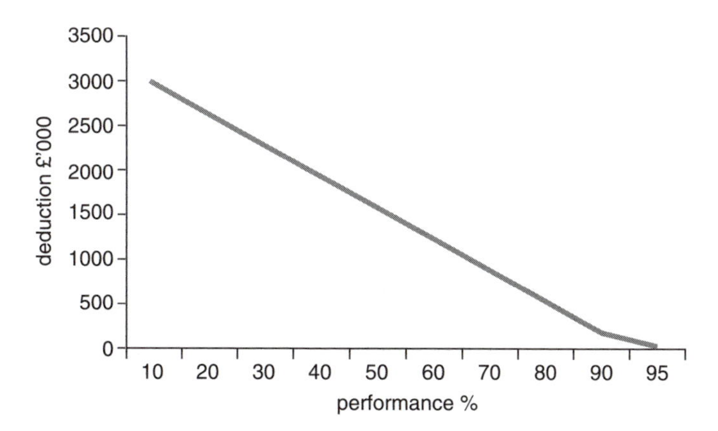

Figure 11.2 Relationship between performance and payment deduction.

during the operation phase of the project. Market testing is used to adjust the payment for the service provision in the operational phase. The aim of this process is to ensure that the value for money continues through time to reflect the true cost of operational service provision. The process of market testing is not universally uniform among PFI/PPP projects. In some projects services are market tested if the project company has incurred more than a pre-specified amount of payment deductions, but otherwise is benchmarked. In others, the client decides whether the services are benchmarked or market tested. But in some FM service agreements the client stipulates that market testing should take place after a specified number of years of the operational phase, in most cases at five-year interval.

The FM service agreement between the client and the project company requires the components of the soft operational service provision to be market tested. The project company should already have priced all FM provision throughout the life-span of the concession at financial close. The exercises take place on pre-specified dates in the contract period. For example, the Norfolk and Norwich Health Care NHS Trust facilities management agreement (2006) states that:

> During the Operational Phase, the Market Testing of each Market Tested Service shall take place so as to have effect on a date (a '*Market Testing Date*') which is, as to the first such Market Testing, on the fifth anniversary of the Phase 1 Completion Date and, as to each subsequent Market Testing, on the fifth anniversary of the date of the last Market Testing.
>
> . . . No later than eight months before the relevant Market Testing Date, the proposal (the '*Draft Market Testing Proposal*') describing in detail for each Market Tested Service or group of Market Tested Services which is to be Market Tested on the relevant Market Testing Date.
>
> . . . If, pursuant to the first Market Testing, the aggregate Market Tested Prices for the Market Tested Services exceed the aggregate Service Fees payable by the Trust to the Project Co in respect of the provision of the Market Tested Services immediately before the first Market Testing Date but do not exceed the aggregate Service Caps in respect of the Market Tested Services immediately before such date, with effect from the first Market Testing Date, the Service Fees for each of the Market Tested Services shall, in each case, be adjusted to reflect the Market Tested Price for that Market Tested Service.

It is anticipated that the project company will invite bids for a soft operational service component from a number of pre-selected providers. The lowest price obtained from the market will become the new price for the service component and the unitary payment will be adjusted accordingly. The new price of the service will be the price that a provider in the market is willing to accept for providing it.

The question here is, is it possible to use both benchmarking and market testing at same time? The answer is that the client should use benchmarking on a yearly basis to test the efficiency of the service provisions and market testing should be used only on a long-term basis and in special circumstances to make sure the service provision still provides value for money to the client. These solutions will provided incentives to the project company and its sub-contractors to implement operational strategies that reduce the cost of providing soft operational services. However, this may conflict with the interests of the lenders. Probably, financiers have no incentive to promote operational strategies that reduce operational costs; due to the fact that such strategies could increase the project risks. Having said this, it is too early to assess the usefulness of market testing as a tool to control value for money in delivering public services because most of the PFI/PPP schemes in UK are still in their early stages.

Refinancing

Refinancing is defined by NAO (2002) as:

> The process by which the terms of the funding put in place at the outset of a PFI contract are later changed during the life of the contract, usually with the aim of creating refinancing benefits for the contractor (bringing forward their returns from a project as a result of changes to a contractor's financing structure).

Refinancing PFI/PPP deals is made possible because of the increased maturity of the PFI market and due to the increased confidence of the lenders on good returns from PFI/PPP schemes. This has led to the creation of a secondary market for PFI/PPP equity, which is allowing investors in projects to sell their shares on to new investors. However, refinancing is an extremely complex issue. Most the PFI/PPP deals signed before 1992 did not include mechanisms whereby the client could share refinancing gains. These contracts are now subject to a voluntary code, which states that the client should generally expect to receive 30 per cent of gains made. However, under the new rules set by the Office for Government Commerce, all PFI/PPP contracts must allow public authorities to receive half of any gains arising from debt refinancing. The new guidance required the private sector to seek clients' approval for all refinancings where the sharing of gains would be applicable. The reasons cited for refinancing include:

- reduction in project risk results in lower finance costs and hence higher rates of return for the private sector shareholders over the life of the project;
- maturity of the PFI/PPP market;

- good financial and operation record of project companies;
- availability of better financial terms (lower debt cover ratios and longer maturity).

In the majority of cases, the main driver is securing further debt against the project. As explained in Chapter 4 the amount of debt which can be raised for a project is directly related to revenue before debt service, divided by debt service. According to Yescombe (2002), reduction in cover ratio will certainly lead to an increase in debt amount. The level of debt can also increase as a result of increases in cash flow or decreases in reserve funds. Yescombe stated that refinancing generally accelerates cash flow to equity investors, but reduces their total cash flow over the project life. The author concluded that most refinancing schemes do not create a saving for investors and the investors' gain is derived from an increase in IRR.

According to Finlay (2003), refinancing of PFI/PPP deals is possible because the lenders might be able to offer better finance terms according to the following scenarios:

- Extending the loan repayment period for the project company. This will permit the project company to pay increased dividends each year to its shareholders, because pressure to pay off the loan has decreased.
- Reducing the interest rate for the project company to reflect the reduced risks.
- Increasing the amount of debt in the project company, creating surplus funds which can be used to accelerate the payment of benefits to shareholders.

Other authors, however, argue that most of PFI/PPP projects are financed on a fixed-rate basis and a general reduction in base interest rates will not automatically lead to refinancing gain. According to Yescombe (2002), the reason for this is that the swap (i.e., exchange of obligations to repay principal interest) provider has to be compensated for termination loss. The NAO (2005a) has also advised procuring authorities to undertake a thorough analysis before agreeing to a refinancing that involves increased levels of private sector debt and higher public sector termination liabilities. The OGC guidance note (HM Treasury 2005) on calculation of the authority's share of a refinancing gain indicated that the client's share of refinancing gain can be paid either as a lump sum or as a reduction in the unitary charge. The lump sum option involves raising a significant amount of new senior debt. This cash is used to prepay shareholders of subordinated debt. The payment distribution should be equal to the amount of new debt which is being raised. The client is entitled to share this gain with the investors and he is entitled to take up to 50 per cent of the first distribution as the lump sum. The reduction in the unitary charge option is mainly due to the fact that the refinancing occurs as a result of an extension of the debt maturity or a reduction in the interest rate. This gain from this refinancing option occurs over longer period. The advice from OGC states: 'The Unitary Charge reduction option will tend to produce a lower Refinancing Gain, as reductions in the Unitary Charge will reduce debt cover ratios, and hence reduce the amount of new debt that can be raised'.

Yescombe (2002) demonstrated that payment of the gain via unitary charge reduction reduces the gain to be shared by about 30 per cent, and on top of this it leaves more of the gain with the private sector. Hence, the HM Treasury note on value for money in refinancing requires procuring departments to consider carefully the risk and value for money implications of a refinancing before giving their consent to a contractor to proceed with the refinancing process, especially where refinancing involves an increase in the client's termination liabilities. According to HM Treasury (2005), the risk from refinancing could result in the following:

- increase in project debt which will have a direct impact on project risk;
- reduction lenders' long-term financial incentives;
- increase in rigidity of the contract and reduction of flexibility for future financial changes.

Granting refinancing is a complex issue. Issues for the client in giving consent to a refinancing, include:

- *Maximum level of senior debt:* this is related to acceptable level of indebtedness of the project company and its financial stability. The client is required to carry out a rigorous evaluation of the maximum acceptable level of senior debt that a project company can withstand without undermining the financial flexibility of the contractor to manage and endure routine and major project risks.
- *Increased termination liabilities*: normally the rules for termination liabilities are set out in the project agreement based on issues such as contractor default, *force majeure* termination and voluntary termination by client. The rule of thumb is that refinancing should not lead to increasing the compensation payable to a defaulting project company. Increased termination liabilities could lead to a reduction in contract flexibility in terms of ability to allow for future changes in policy affecting the services provided under the contract and operational flexibility to cope with the rapid change in service demand. The Treasury advised that the breakage cost (as a result of early payment of debt) and profile of termination liability (time-dependent profile of termination liabilities) need to be considered if an increase in termination liabilities is necessary. In any case, clients are unlikely to agree to a refinancing that increases its termination liabilities.
- *Contract amendments*: related to amendments to the contract between the purchaser and provider and to the direct agreement between the client and

senior lenders. These contract changes are necessary to provide payment to the client from the share of the refinancing gain.

- *Changes to the unitary charge profile or indexation*: these aspects of the PFI/PPP deal are normally fixed at the financial close. Hence, any change on these parameters may lead to revaluating value for money aspects of the whole deal. The rule of thumb stipulated by the UK Treasury is that any changes to the profile of unitary charge payments and indexation regime are not value for money unless proved otherwise.
- *Contract extensions*: to produce refinancing gain or reducing unitary payment, sometimes it is necessary to extend the concession agreement. This will enable the provider's debt to be amortised over a longer period, leading to a decrease in debt servicing costs. The guidelines require that if contract extension cannot be justified as value for money on a stand-alone basis, it should not form part of a refinancing proposal.

Condition surveys for maintenance management

Building components deteriorate over time. Keeping these components in good condition during their entire service life is a crucial aspect of maintenance management. A maintenance strategy that is based on a rational assessment of reliability and predictability of the service life of building components and on real-time knowledge of the condition of assets components will help to ensure effective management of life cycle funds. These issues have led many authors to advocate the use of condition surveys as a strategy for maintenance management. Condition surveys are defined by Dann and Worthing (2005) as: 'A systematic survey of the fabric of building, in order to produce accurate information of the condition, and an assessment of the extent and timing of future maintenance, repairs and replacement.'

The aims of condition surveys are listed as follows, according to Then (1995) and Dann and Worthing (2005):

- Collect data on general condition of all parts of the building.
- Reconcile progress of repairs and planned maintenance programmes.
- Identify the need for further repair.
- Identify emergency repair.
- Develop programmed maintenance strategies.
- Develop responsive maintenance strategies, prioritise maintenance and repair programmes.

Then (1995) explained that building condition surveys are essential in prioritising maintenance programmes. The intensity of maintenance work is directly related to the level of difference between the current condition of building fabric and the specified or desired condition at any stage of the service life of building assets.

Alni *et al.* (2000) developed a model that relies on the relationship between the level of defects and time over the intended service life of building fabric. The authors proposed the use of cost/time curve for the management of building maintenance and repair. Noortwijk and Frangopol (2004) propose the use of condition-based and reliability-based maintenance optimisation models for the life cycle maintenance of infrastructures. For PFI/PPP schemes, the building fabric has been pre-determined at the design stage, and the building is already some way through its service life, and therefore planning will focus on assessing the residual service life of the components and optimising planned maintenance and costs of replacements. Hence, probably it is wise for PFI/PPP stakeholders to incorporate such strategies in PFI/PPP deals.

It is disappointing to find that little information, if not none at all, is provided in project agreements or for that matter in the facility management agreements, on building condition surveys and appraisal of maintenance.

The client should specify in the FM contract or project agreement that either the FM operator or project company should report on the condition of the building at least once a year. This will reassure the client that the quality of building is being kept to the standard required. This will enable the FM operator and project company to plan more accurately for life cycle expenditure and develop strategies for making decisions on maintenance prioritisation, i.e., postpone or move forward maintenance and replacement of components. The client should put in place a provision to commission independent condition surveys on the building if he is not satisfied with the quality of maintenance regimes.

Controlling maintenance and repair cost

The implementation of whole life cycle decisions could result in a marginal capital increase, but may result in operation cost efficiency of the PFI/PPP schemes. For PFI/PPP schemes, building fabric have been pre-determined at the design stage, and the building is already some way through its service life, and therefore planning will focus on assessing the residual service life of the components and optimising planned maintenance and costs of replacements. Mills (1994) highlights the importance of design decisions upon the likely maintenance costs that will arise in the future. The design decisions made will dictate the likelihood of component failure and how the incidence of use will place particular demands upon the building structure and fabric. Mills indicates that there are five situations in which the building design team can influence the maintenance costs. These are, during the design brief; during the choice of materials; during the detailing; access with minimum inconvenience; and the prompt replacement of components at the end of their service life. Leifer (1997), in a study of the economics of building ownership, draws attention to the fact that maintenance costs become increasingly onerous as the wear and tear on the fabric and structure (including services) increases.

Flanagan and Norman (1983) who compiled a criterion to facilitate the correct choice of action, highlight the following factors that can affect the maintenance of buildings:

- rate of deterioration of the building and/or element;
- the cost of different types of repair;
- the disruption and disturbance to the building occupants and time required for repair;
- the relationship between the physical life of the repair and the required physical, functional and economic life of the building.

Holmes (1994) draws a distinction between the two types of maintenance, response (normal and emergency) and programmed maintenance (cyclical and preventative). Response maintenance is usually initiated by the occupier and will be categorised by a number of factors such as type of work required, nature of fault and seriousness. Cyclical maintenance is that which is carried out irrespective of the condition of the elements, whereas preventative maintenance is condition dependent. The NHS Estates (2004) *Design Development Protocol* outlines the role and responsibilities of the PFI/PPP contractor regarding the management of maintenance as:

> Upon completion of the design and construction contract, the PFI contractor will be responsible for the maintenance and replacement of all Group 1 and maintenance of Group 2 fixtures and fittings/equipment for the whole of the concession period. For both groups, this will include any making good/modifications/damage to the building fabric and/or engineering services during equipment decommissioning and replacement. When Group 2 equipment ceases to be maintainable, the Trust will be responsible for the supply costs only of life-cycle replacement items of Group 2 equipment over the concession period and the PFI contractor will be responsible for installation and maintenance thereafter. The disposal of replaced Group 2 equipment will be the contractor's responsibility, unless the trust elects otherwise.

Before the start of the operational phase, the client stipulates that the following aspects of the maintenance preparatory work should be put in place (Norfolk and Norwich Hospital 2006):

- adequate staff reserve in place;
- work schedules available for all areas;
- planned preventative maintenance system operational;
- maintenance response system in place;
- management system for maintenance and engineering activities available to meet the requirements of this schedule.

The process that determines maintenance costs is included in all PFI/PPP project agreements. Generally, the project company submits to the client a yearly programme for the performance of all planned maintenance, repair, renewal or replacement relating to the maintenance of buildings and engineering services. Normally, the annual programme identifies aspects of the facility, including external works, which are concerned with such planned maintenance, repair and renewal/replacement. The period during which these maintenance activities are taking place is identified and agreed with the client. The timing of the maintenance work may have a knock-on effect on the use of some parts of the facility. The unavailability for use is normally documented in the project agreement and dealt with accordingly. The annual programme and timing of maintenance are, in general, subject to agreement between the client and FM service provider. It is common practice that the project company bears all consequences of deferring maintenance, repair, renewal and replacement of building and engineering service components. Deferral of maintenance programmes could have a materially adverse effect on increasing the likelihood of the occurrence of unavailability and may also have an impact upon the cost or availability of building insurance. What is important here is that the FM operator must make sure that maintenance and replacement programmes must not compromise the operational expenditure certainty of the project.

The fee for the maintenance of buildings and engineering services is a part of the usage fee or unitary charge. However, the costs used by the project company in carrying out planned preventative maintenance to the extent that such costs are below a threshold value are dealt with as reimbursable expenses.

The budget for planned maintenance, repair, renewal or replacement relating to the maintenance of buildings and engineering services should be developed prior to financial close and is embedded in the service agreement. In the majority of projects, this is referred to as the life cycle fund. The agreement on who controls this fund varies significantly from project to project. In some PFI/PPP contracts, the life cycle fund and its management are included as part the FM agreement. In some cases the lenders may demand that the project company should retain control and management of life cycle funds. This stipulation is mainly due to the fact that if FM operator defaults, the life cycle funds will be unaffected and would lead to a reduction of the overall project risk. There are other scenarios where the life cycle funds are controlled by both the lenders and the project company. In such agreements, the funds are released by lenders and the project company based on recommendations from the facility management contractor. The problem with this scenario is that if life cycle expenditure has peaks, this could prevent the project company meeting its other liabilities from unitary payment on these periods where major maintenance and replacement peak. This issue is generally dealt with through the selection of durable building components that will lead to smoothing the life cycle maintenance expenditure. It is normal practice that whoever controls the life cycle fund is entitled to any surplus in the fund at the end of the concession agreement. This should motivate the life cycle fund control to select a maintenance life cycle solution that minimises life cycle expenditures and maximises profits. Hence, it is in the purchaser's interest to insist that the provider and his subcontractors must implement design, construction and operation solutions that lead to enhanced value for money to every stakeholder involved in the delivery of the PFI/PPP deal.

Controlling operational costs

As stated above, the guidelines for service performance measurement are normally dictated by clauses of the project agreement. The PFI/PPP contracts are based on the paradigm of 'self-monitoring' in that the PFI provider is responsible for providing and reporting quality service aspects. However, the client has his own team to lead supervision and monitoring of the provider's performance in terms of meeting the required standards for the availability of the PFI/PPP facilities and the delivery of FM services, to confirm satisfactory delivery of the contract obligations. This information is then used as the basis for approving regular contract payments to the PFI provider. According to Edwards *et al.* (2004): 'Monitoring is an accountability procedure intended to ensure that the public sector agency discharges its responsibilities in relation to the procurement and availability of services and that payment is made properly in line with the contract.'

In some cases, both representatives from client and service provider organisations monitor the performance of each contract service. Some service agreements stipulate that for the purpose of monitoring, the client, within a specified number of days of every contract month selects a specified percentage (e.g., 15 per cent) of the total activity outputs for each service on a random basis for sample monitoring the performance of the project company. Audit Scotland (2002) describes the performance measurement in PFI schools as:

> The foundation for service monitoring and reporting is the PFI provider's helpdesk service, which receives progresses and reports all service requests from schools and any problems. Failures to provide service must be remedied within specified periods and failures will lead to payment deductions, and ultimately for serious non-performance may lead to contract suspension or termination.

In most PFI/PPP schemes, monitoring and reporting is established along the following principles:

- *Weekly performance monitoring*: the aims of this process are to co-ordinate and communicate all day-to-day operational activities.
- *Monthly performance reports by contractor*: the aim of this report is to inform the stakeholders about the detail standards and levels of performance

actually achieved during the month concerned across all FM service provision.

- *Monthly stakeholders performance monitoring*: the aim of this process is to review monthly performance reports, agree payment deductions and address all performance disputes.
- *Strategic performance review*: the aim of this mechanism is to review contractual matters and efficiency of delivering the required FM services. Issues relating to variations to the project agreement, performance measurement, monitoring, etc. are also addressed at this level.

Performance measurement can vary considerably from scheme to scheme. The most common aspect of services agreements is that all of them specify a minimum service threshold for the operator to be paid in full for particular services. In the majority of PFI/PPP contracts, the performance of FM services is measured according to a percentage scale. The FM performance is expressed as a percentage of achieved performance standard. The aggregate performance is computed on the basis of each of the FM service types. For example, the service agreement for Norfolk and Norwich Hospital (2006) states that:

- The aggregate performance of each FM service element should be at least 96 per cent or above, the required performance standard for paying the contractor in full for particular service.
- Below 96 per cent, the contractor may become liable for payment.
- If the performance percentage falls below 95 per cent for 12 successive contract months, the client increases the level of its monitoring of the contractor over and above the normal level, until the contractor attains the required performance standard for that service during three successive contract months.
- If the performance percentage falls below 90 per cent for three successive contract months, increased monitoring will be triggered as described above.
- If the performance percentage falls below 75 per cent for six consecutive months for one or more of the services, the client is able to issue a termination notice for that FM area.
- If the performance percentage falls below 85 per cent for 12 consecutive months for one or more of the services, the client is able to issue a termination notice for that FM area.

An example of FM service elements and their associated percentage weights is presented in Table 11.1. The key indicators shown in the table are normally used as the basis for scoring FM service provision. NAO (2005a) found the weighting used in the measurement of FM services delivery is inappropriate and sometimes is not consistent across all measured services.

Table 11.1 An example of FM services performance measurement

FM element	*Performance indicators*	*Percentage weight*	*Standard performance*
Grounds	Ground maintenance	10	The aggregate performance of each FM service element should be at least 96% or above, the required performance standard for paying the contractor in full for particular service
	Maintain access points	8	
	Periodic planting	8	
	Staff training	8	
	Good, safe horticultural practice	8	Below 96% the contractor may become liable for payment
	Staff attitudes/behaviour/uniforms	8	
	Environmental policy review and update	10	If the performance percentage falls below 95% for 12 successive contract months, the client increases the level of its monitoring of the contractor over and above the normal level, until the contractors attains the required performance standard for that service during three successive contract months
	Snow and ice clearance	10	
	Maintain safe environment	5	
	Response to routine repair	5	
Security	Regular/comprehensive staff training	10	
	Staff uniforms clean/appropriate, staff attitudes/behaviour	4	If the performance percentage falls below 90% for 3 successive contract months, increased monitoring will be triggered as described above
	Adequate levels of trained/competent managers/supervisors	6	
	Up-to-date records of all complaints received and action taken	6	If the performance percentage falls below 75% for 6 consecutive months for one or more of the services, the client is able to issue a termination notice for that FM area
	Adequate audited quality control system and internal monitoring	10	If the performance percentage falls below 85% for 12 consecutive months for one or more of the services, the client is able to issue a termination notice for that FM area
	Patrolling routes/work schedules regularly reviewed/updated	12	
	Compliance with Trust policies/procedures and statutory requirements	15	
	Allocation of staff to ensure adequate response to incidents	15	
	Correct use of equipment and recording requirements	12	

Maintenance of buildings and engineering services	Information/liaison requirements of the Trust	10
	Response to emergency repair	10
	Contractor ensures safety and cleans up after job	7
	Staff qualifications – statutory and NHS	7
	Staff attitude/behaviour/uniforms	7
	Response to safety action bulletins, hazard warning notices, etc.	7
	Complaints received and action taken	7
	Response to urgent repair	8
	Maintain internal environment	4
	Comply with planned maintenance	2
	Response to routine repair	1

Source: Norfolk and Norwich Hospital NHS Trust

According to NAO (2005a), the aggregate performance of FM services each month is developed based on:

- monthly summary of all comments/complaints received;
- monitoring reports from FM provider;
- joint inspections and auditing.

Performance reports are normally generated at the end of every contract month. The performance reports are generated from a combination of data as listed above. These reports set out the standard achieved in respect of FM services during the contract month for each service. The client has the right to audit, from time to time, data provided by the provider in relation of any activity output assessment.

There are regular meetings, usually monthly, between the PFI provider, the client and individual providers responsible for delivering services. These meetings are used to discuss the scores to be allocated to the FM services and any deductions for non-performance. Once the performance for the month is decided, the service provider within a specified number of days of receipt of performance report must inform the client of any disputed performance measurement. Any concerns are normally resolved either through further negotiation or through disputes resolution process before payment is made (further details on dispute resolution can be found in Chapter 12).

In some service contracts, where the provider exceeds the standard performance in respect of any service, the amount of such excess might be used to compensate for a failure in that particular service within a specified period (e.g., within 6 months). NAO (2005a) argued that if the contractor accumulates enough credits, he could lower FM service standard, as he will not suffer payment penalties. Hence, this practice is no longer recommended.

Under all PFI/PPP contracts, the client is entitled to make payment deductions for unsatisfactory service performance. But the deduction regime can vary significantly between PFI/PPP deals. As an example the Norfolk and Norwich Hospital Services agreement deduction from the amount of the service fee payable is determined in accordance with the following formula:

$$Y_d = Y_{sf}(P_r - P_a)$$

Where:

Y_d = the amount of the relevant deduction

Y_{sf} = the service fee payable in respect of a particular for the relevant contract month

P_r = the required performance standard for the service concerned expressed as a percentage

P_a = the achieved performance standard for the service concerned expressed as a percentage.

Results generated from the above formula are plotted in Figure 11.2. The figure shows that this particular payment deduction system is robustly developed to correctly provide for the amount of the deduction to increase in line with the seriousness of the service performance failure and for the possibility that deductions in respect of a particular service could be made up of the full amount of contractor's charges for the service in question in any month. As explained above deduction does not occur automatically. Normally various stages have to occur before there is a reduction in the fees payable. A deductive performance regime like the one presented here will provide incentives to the FM contractor to devote more resources to deliver the FM services at the required standard level.

Summary

This chapter introduced the process of monitoring costs at the operation stage of PFI/PPP projects from the point of view of clients and providers. The process of setting control benchmarks is also introduced. Measurement of the performance of PFI/PPP schemes is complicated and maybe dependent on many outcomes of a different nature. The problem is further complicated by lack of standards against which to benchmark. The processes and principles by which technical and financial performance measurement systems would operate to measure the delivery of facilities management services are explained. Normally the project agreement sets out the mechanisms for this process. It is disappointing to find that little information is provided in project agreements or for that matter in the facility management agreements, on building condition surveys and appraisal of maintenance. However, the process that determines operational and maintenance costs is included in all PFI/PPP project agreements.

Part V

Risk in costing PFI/PPP projects

Chapter 12

Risk in cost planning of PFI/PPP projects

Introduction

A distinction can be made between cost decision-making under uncertainty and under risk. Risk concerns situations where objective probability distributions of expected likelihood of occurrence can be constructed. Uncertainty, meanwhile, is associated with only partial knowledge, the results of which cannot be determined objectively. A probability distribution, therefore, is not available. For simplicity in this chapter, unless stated otherwise, we will use the two terms interchangeably.

Cost decisions of PFI/PPP projects are complex, and usually have many significant factors affecting the ultimate cost decisions. PFI/PPP cost decisions generally have multiple objectives and alternatives, long-term impacts, multiple constituencies within the PFI/PPP procurement system, involve multiple disciplines and multiple decision-makers, and always involve various degrees of risk and uncertainty. In PFI/PPP cost decisions, there will be numerous variables influencing the risk, but the impacts are not explicitly included in the mathematical decision-making models. Whole life cost decisions and their associate risks must be quantified, analysed and presented as part of the strategic decision-making process in today's business environment (Boussabaine and Kirkham 2004). Hence, the aim of this chapter is to tackle some of the issues associated with the risk of costing PFI/PPP schemes. This chapter presents a brief introduction to the methods of risk assessment and explains the risk factors facing the cost analyst in pricing PFI/PPP capital and operational costs. Effect of risks on PFI/PPP project prices are also discussed in parallel with risks that may lead to capital cost variation. The issues of operational and maintenance pricing risks are also introduced in this chapter. Strategies for risk are introduced and exemplified.

The life cycle of risk management in PFI/PPP projects

Risk management is an integral part of the PFI's procurement processes and procedures. The whole concept of PFI/PPP arrangement is based on an appropriate and clear allocation of risks and responsibilities, thereby delivering value for money to the client through minimising the potential for future disputes and difficulties of cost overruns. The risk management process in PFI/PPP projects is aimed at achieving the following objectives:

- to demonstrate value for money for decision-makers;
- to identify all major risks relevant to PFI/PPP procurement systems;
- to increase understanding of risk allocation in PFI/PPP contracting systems;
- to deliver a robust financial and contractual structure for the project;
- to create a risk management process during procurement and operation of the concession agreement.

Risk management is an ongoing process that continues throughout the life cycle of the concession period of PFI/PPP projects and occurs in the stages as outlined in Figure 12.1.

- *Risk identification*: the process of identifying and extracting all possible risks that are associated with PFI/PPP schemes. The meticulousness of this

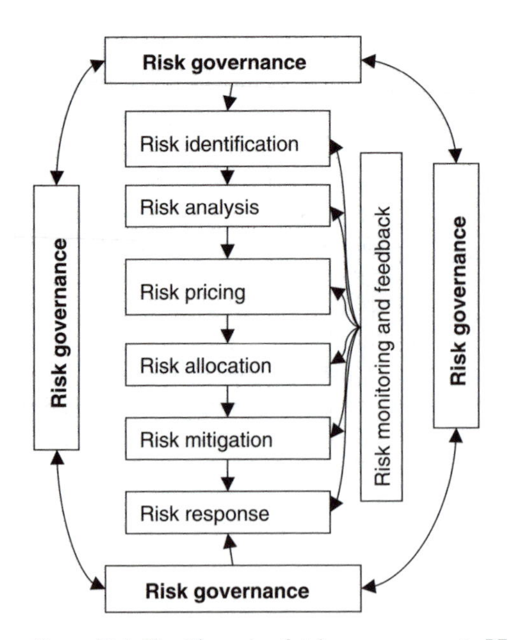

Figure 12.1 The life cycle of risk management in PFI/PPP projects.

process has a direct impact on the subsequent risk cycle process throughout the life span of the PFI/PPP deal. The output of risk identification is normally produced on a risk register covering key risk areas and individual risks within these areas of the PFI/PPP schemes (see Chapter 13 for a typical risk register).

- *Risk analysis*: the process uses analytical methods to determine the likelihood of identified risks occurring and the magnitude of their consequences if they do materialise. There are several options available to the analyst to do this (see the next section). PFI/PPP decisions are usually assessed against the risk register, assessing the impact, probability and exposure using an analytical method. The overall exposure to risk is arrived at through the product of the impact of risks and likelihood of them occurring.
- *Risk pricing*: the process of putting a value to each of the risks using estimates of probability, impact and timing.
- *Risk allocation*: the process of allocating responsibility for dealing with the consequences of each risk to one of the parties of the PFI/PPP deal, or agreeing to deal with the risk through a specified mechanism which may involve sharing the risk.
- *Risk mitigation*: the process used to reduce the likelihood of the risk occurring and the degree of its consequences for a particular stakeholder. Methods used for risk mitigation include: risk reduction, risk avoidance/elimination, risk transfer and risk retention/absorption. The degree of significance given to any particular risk varies from project to project and from stakeholder to stakeholder.
- *Risk responses*: the process of developing strategies to deal with anticipated risk threats or consequences. Universal rules such as risk avoidance, risk reduction, risk absorption and transfer are used for this purpose.
- *Risk monitoring and feedback*: the process of monitoring and reviewing identified risks and new risks throughout the life cycle of the agreement.
- *Risk governance*: the process of managing all the risks identified in the risk register for the PFI/PPP deal. The process should address the issue of who is responsible for the implementation, control and monitoring of the actual state of risk responses and management strategies throughout the life cycle of the PFI/PPP contract.

Methods for risk assessment

A large number of techniques have been used to assess uncertainty with regard to costing decisions. When cost analysts are faced with pricing a project or part of a scheme under uncertain conditions, they are mostly concerned about how to work out a price that will reflect the magnitude of the anticipated risks. The outcome of such difficult costing decisions might be less favourable than what is acceptable. The techniques that can be used in cost risk assessment decision-making are classified by Boussabaine and Kirkham (2004) as deterministic, qualitative and quantitative.

Deterministic techniques

Deterministic methods measure the impact on project outcomes of changing one uncertain key value or a combination of values at a time. These methods are based on techniques such as sensitivity and breakeven analysis methods that are easy to use, understand and require no additional methods of computation beyond the ones used in normal cost analysis. Deterministic models compute prices and costs as a single-point estimate. What is usually missing from these models is a calculation of the probability of the estimate occurring along with an estimate of the degree of risk involved in each of the estimated prices. Several risk analysis methods are classified under this approach including:

- conservative benefit and cost estimating;
- breakeven analysis;
- risk-adjusted discount rate;
- certainty equivalent technique;
- sensitivity analysis;
- variance and standard deviation;
- net present value.

Qualitative techniques

Qualitative risk analysis is where the likelihood or the magnitude of the consequences of an event or occurrence is expressed in qualitative terms as opposed to quantitative risk analysis where the probability or frequency of the outcomes can be estimated and the magnitude of the consequences is quantified. The following techniques are utilised in this method:

- risk matrices;
- risk registers;
- influencing diagrams;
- SWOT analysis;
- risk scoring;
- brainstorming sessions;
- likelihood/consequence assessment.

According to Boussabaine and Kirkham (2004), qualitative risk assessment methods are used as follows:

- as an initial screening activity to identify risks that require more detailed analysis;

- where the level of risk does not justify the time and effort required for a more complete analysis;
- where the numerical data are inadequate for a quantitative analysis.

Probably the commonest method in the procurement of PFI/PPP schemes is risk matrices. At least, it is a minimum requirement that all procuring authorities must develop a risk matrix for their PFI/PPP projects before approval.

Quantitative techniques

Quantitative approaches are based on the assumption that no single figure can adequately represent the full range of possible alternative outcomes of a risky investment. Rather, a large number of alternative outcomes must be considered and each possibility must be accompanied by an associated probability of a probability distribution. Statistical analysis can be performed to measure the degree of risk. In the case of the deterministic approach, the analyst determines the degree of risk on a subjective basis. Quantitative risk methods use the following techniques:

- input estimates using probability distribution;
- mean-variance criterion coefficient of variation;
- decision tree analysis;
- simulation (MC/LHS);
- mathematical/analytical technique;
- artificial intelligence;
- fuzzy sets theory;
- event trees (quantitative);
- complexity tools.

As explained above, the input variables in pricing decision problems are dealt with deterministically and probabilistically using techniques such as Monte Carlo simulation. While these techniques do provide a basis for making decisions under uncertainty, they often do not account for many of the parameters that may affect the outcome of decisions. These methods are based on top-down approaches and are unable to take into account the low-level interactions between risk variables that lead to losses. Mechanisms underlying risk may exist at a lower level in the hierarchy (Johnson 2005). Recent approaches in dealing with the impact of parameter uncertainty also include fuzzy set theory and artificial intelligence. The pricing of PFI/PPP, particularly the forecasts of maintenance and operational costs can be regarded as highly uncertain (due to a time horizon of 30 years ahead). Fuzzy set theory can incorporate imprecision and subjectivity into the model formulation and solution process. Fuzzy set theory in decision-making can be viewed as a generalised sensitivity approach to the precision with which input variables are known to the decision-maker.

These approaches can be extended to model risk factors in PFI/PPP procurement. Other methods such as neural networks, agents-based approach, cellular automata, game theory and chaos theory could be extended to model risk in cost planning.

The use of risk assessment techniques, in the pricing of PFI/PPP deals, will allow both the client and providers to directly address the probability features of investment (in maintenance, operation, energy programmes and capital programmes of projects) by conducting a large number of what-ifs on each uncertain decision variable in the pricing process. This type of analysis will allow the cost analyst to address the following questions.

- What are the most likely capital, maintenance and operating costs?
- How likely is the baseline cost to be overrun?
- What is the whole life cost risk exposure?
- What is the likelihood of corporate loss of revenue, business failure, etc.?
- Where is the risk in pricing PFI/PPP schemes?
- What is the risk of accepting or rejecting alternative pricing decisions?
- How likely is the selected decision/price optimal?
- What are the risks associated with whole life costs?

None of the above techniques can be applied to every situation. What is best depends on the relative size of the project, the availability of data and resources, computational aids and skills, and cost analyst understanding of the technique being applied. That is why we always advocate the use of a combination of techniques to price risk in cost planning of construction projects.

Uncertainty in costing PFI/PPP projects

Estimating future costs involves uncertainty in developing long-term cost assumptions. If there is substantial uncertainty concerning cost and time information, cost plans, especially whole life costs, may have little value for decision-making. Time and cost risks occur when the analyst makes errors in assumptions or on the actual pricing of any aspect of the PFI/PPP deal. Errors in developing costing assumption are very common in the construction industry. Probably one might argue that the reason why most of the public sector projects cost outturn is higher than estimated is mainly due to errors in project estimation as a result of making inaccurate judgements and assumptions.

The main causes of uncertainty in cost planning of PFI/PPP projects can be attributed to the following issues:

- difficulty in achieving predictability of long-term future costs;
- complexity of the procurement system and its ability to deal with rapid changes in product and process. It is a known fact that building and operating cost outturns are highly sensitive to changes in initial assumptions.

- incomplete or inaccurate assumptions due to deficiencies in data and simplification of costing assumptions by assuming normality of events, rather than chaotic conditions;
- incomplete methodology, mainly due to use of models that do not include all relevant costing parameters;
- lack of skills and experienced cost analysts to deal with complex costing processes.

The process of estimating or pricing PFI/PPP contracts is confounded by imprecise information about baseline costs, especially life cycle costs. Boussabaine and Kirkham (2004) have advocated that the uncertainty in baselines prices of building cost centres should be dealt with by taking more than one estimate. The philosophy here is that whole life costs should not be priced as a single value, but as ranges based on the level of existing uncertainty. The higher the uncertainty, the lower the degree of confidence in estimated prices. The lower the degree of confidence, the higher the estimated range should be. The authors suggest the use of estimating strategies that give a range of prices from low, through mid, to high prices. This approach provides a sensitivity analysis for costing PFI/PPP schemes at both basic and project levels. In this way the cost analyst and decision-makers will have a full set of plausible prices on which to base their costing decisions.

The following strategies could be utilised by stakeholders to minimise the impact of errors in pricing PFI/PPP deals:

- develop accurate pricing methods using simulation methods;
- allow an appropriate risk reserve;
- transfer risks by subcontracting most of the work through fixed price contracts, based on firm sub-contract quotations;
- optimise project price and margins in relation to the level of risk on PFI/PPP schemes;
- develop cost sign-off gateways;
- provide peer review panel/workshops to evaluate estimates/forecasts and the underlying assumptions and methodologies;
- minimise hidden costs such overheads and administration costs.

Impact of risk on PFI/PPP project costs

The level at which risk is priced and the magnitude of risks transferred to the private sector will have a significant impact on the cost of the PFI/PPP deals as well as on the value for money analysis and on the section of the optimum investment options. Figure 12.2 shows that, in each PFI deal, the NPV cost of a public sector comparator is lower than a PFI deal before risk transfer is added to the public funded projects. Figure 12.2 clearly demonstrated that only after risk price is added to project cost is the NPV cost of the PFI deal marginally lower

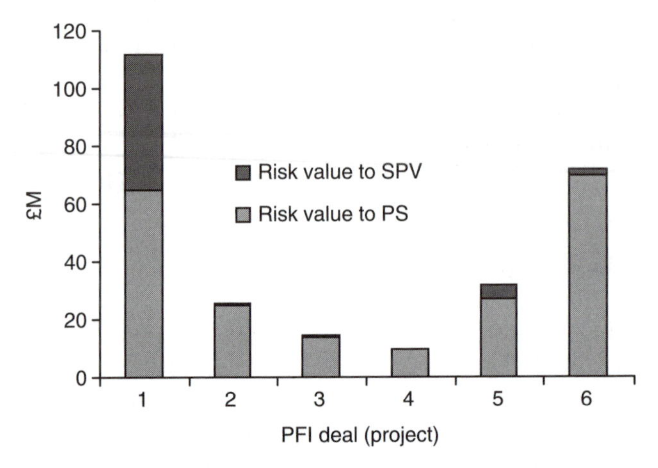

Figure 12.2 Impact of risk on PFI deals.

than the public sector option. Only after risk transfer, the difference between the PFI/PPP and the public sector comparator in all cases is marginal (Pollock *et al.* 2002). It also shows that there is a wide variation in the value of risk transferred to the private sector. One may argue that at least in these projects the price of risk transferred to the private sector might have played a major part, if not be the decisive factor, in selecting the PFI deal in preference to the public sector option. This view is shared by many authors. For example, Shaoul (2005) stated:

> At the very least, this suggests that the risk assessment methodology is somewhat arbitrary and that the value of the risk to be transferred was calculated in such a way as to close the gap between the PSC and PFI. But more importantly, even after including risk, the margin of difference between public and private is tiny, typically less than 1 per cent of the total costs, and not enough to provide a criterion upon which to base such an important financial decision.

Pollock *et al.* (2002) analysed six health sector projects. The authors found that in all of these six cases the margin of advantage was in favour of the PSC, before any adjustment for risk. The evidence from the projects shown in Figure 12.2 indicates that cost savings from risk transfer are highly questionable. The problem here is that, on one hand, the purchaser (client), more or less, guarantees annual payments to the private sector, especially the lenders, which amounts to a reduction of risk to the private sector, and, on the other hand, he transfers risk to the private sector at a higher price than it is valued by the private sector. Hence, the private sector might receive extra value from the PFI/PPP deals, through risks it does not bear. Probably this will amount to the double cost of risk value to the client. The way the cost of risks is computed will also have an impact on the cost of PFI/PPP deals. This is particularly important in the selection of the discount rate to be used in the calculation of cost of risks. Shaoul (2005) argues that the computation of costs of risk should not be based on discounting for time value of money that reflects uncertainty, but it should be based on an appropriate discount rate that reflects the social time preference value of money (the method assumes that different costs and benefits occur at different points in time).

Since the essence of PFI/PPP lies in the transfer of risk to the private sector, the impact of risks value on both capital and operational expenditures should be subject to rigorous analysis to determine the relationship between project cost changes and the level of risk value transferred to the private sector. This should be an essential or integral part of affordability and value for money analysis. In carrying out this process, the cost analyst should not inflate the value of risks to the public sector. The assumptions for pricing risks should be realistic and comparable to the private sector.

Cost analysts tend to use sensitivity analysis to determine the impact of risk change on project costs. It is important here that the analyst should understand that sensitivity analysis does not measure or adjust for risk. It only enables the cost analyst to identify critical risk variables that might have an impact on the project cost. The issue here is the determination of the optimal value or cost attributed to each risk factor. For this purpose, the analyst ought to consider using simulation techniques. For complex projects, such as PFI/PPP, that are based mainly on risk valuation and transfer from the public to the private sector, we ought to use robust methods for risk pricing and analysis. This is indispensable in achieving optimum risk transfer and value for money.

Risk of capital cost variation

It is common knowledge that capital cost growth is related to cost overruns, unit cost escalation and project scope escalation. It is also well recognised that the scope and complexity of some PFI/PPP schemes, in the UK, increase substantially between the outline business case and the financial close stages. This has led to large increases in capital costs, to the extent that many authors have questioned the validity of affordability analysis results. One might argue such a rise in capital costs can erode the real value of money aspects that are projected at the original project budget estimation.

It is well known that cost estimation for both capital and operational expenditures of PFI/PPP construction schemes is particularly complex and uncertain, often leading to either considerable risk cost transferred to the private sector or cost overruns. The reason cited for such problems is the long-term nature of concession agreements. Normally capital and operational costs are developed based on previous experience and they are usually based on a single cost estimate. One expects that there should be a trade-off between certainty of capital and operational expenditures and the amount of risk that is transferred to the private sector. At the moment this relationship is unknown. However, the evidence from existing schemes, as demonstrated above, shows that there is little correlation between the risk value and the cost outturn of PFI/PPP projects. That is, the value of risk to be transferred to the private sector is disproportional to capital expenditures.

One of the most important steps in evaluating PFI/PPP projects is the estimation and projection of both capital and operational expenditures. These two costs are essential to the economic and financial viability of PFI/PPP schemes. Capital costing decisions are made, most times, under conditions of uncertainty, and as a result the cost estimates used in PFI/PPP projects appraisals are themselves uncertain and present an element of risk, i.e., cost overruns. This in itself creates uncertainty and, as a result, a level of risk in the estimated cash flows may not materialise and the timing of the cash flow may also vary considerably from the original estimation (Lefley 1997). Humphreys (1991) argued that: 'A cost for a venture has a greater chance of being fifty percent higher than the estimated cost than it does of being fifty percent lower.'

This uncertainty originates from the difficulty of forecasting, with great certainty, how much it will cost to build and operate PFI/PPP schemes over a long period. The evidence of this is shown in Figure 12.3. Figure 12.3 shows how estimated capital costs vary considerably between the SOC and the FBC stages. This confirms the view that uncertainty in estimating capital costs increases with time, the longer the project duration (time here is used as a measure of project complexity), the greater the difficulty in predicting the project cost outturns. This uncertainty in itself creates a risk that the project cost overruns are more likely to occur. The unpredictable factors that contribute to capital cost variations are usually described as risks. In PFI/PPP schemes, these risks,

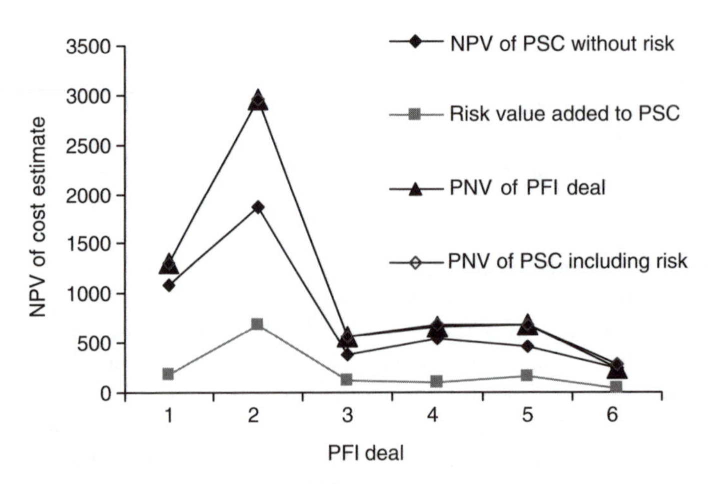

Figure 12.3 Impact of risk on PSC outcome.

from the client point of view, mainly occur at the pre-financial close stage. The UK Treasury guidelines have attributed the cost of capital cost variations to the following causes (HM Treasury 2004b):

- poor definition of the scope and objectives of projects in the business case, due to poor identification of stakeholder requirements, resulting in the omission of costs during project costing;
- poor management of projects during implementation, so that schedules are not adhered to and risks are not mitigated.

There are several methods available to the cost analyst when taking into account risk in evaluating or estimating capital costs. These methods vary in sophistication and complexity, ranging from subjective adjustment of estimates to simulation based on probabilistic distributions of each capital cost element of the project. Methods that are based on simulation are generally better for evaluating capital cost to risk variations. Further details can be found in Boussabaine and Kirkham (2004). The cost analysts should also bear in mind that risk of capital expenditures variation could be reduced significantly through careful cost planning at the early stages of project development (see Chapters 6 and 7) and through comparator cost studies.

Dealing with operational cost risk

Increases in projected operating costs can be critical to the economic viability of PFI/PPP projects. The importance of the risk of variable operating costs is well documented in the public sector facilities and to some extent in some of the private sector utilities. PFI/PPP contracts involve an element of capital risk sharing, with a significant portion of that risk to be taken by the private service provider. In most cases the operational cost risk is entirely transferred to the project company. The project company meet operating costs from charges levied on clients who use the facility. Charges are determined as part of the PFI/PPP process and include the price of the risks taken by the project company.

The majority of PFI/PPP contracts have over 30 years of service contract. A concession agreement of this length without a doubt raises significant questions about operational costs and their validity. The overriding principle in dealing with long-term operational cost risks is to use a combination of strategies including contractual renewal, adjustment of risks, and reward through market testing and negotiations.

Financial lenders will be concerned about operational cost inefficiency and in some circumstances may even choose to intervene where profitability is at risk, as a result of high operational costs. The service provider is normally concerned about allocation of operational risks and how the purchaser seeks to manage operational risks and non-performance penalties.

The key area of risk that is associated with the operational cost risk analysis varies considerably among projects, but in general includes facilities management and availability risks. The quantification of the facilities management risk factors is usually assessed by identifying the likely cost increase, either as a percentage or a lump sum, as having a particular probability of occurrence, thus producing an overall price or value for each risk factor. Sensitivity analysis, varying the level of probability of occurrence and value of risk, is then used to adjust the operational costs to take into consideration the risk value. Availability risks are normally assessed using the payment mechanism. The payment formula is used to simulate a number of potential performance scenarios.

In PFI/PPP procurement systems, the client transfers the risk costs of all operational services to the private sector. The operational costs include the cost of rates, cleaning, energy, portering, security, etc., that are necessary for delivering public services. Operation cost risks are directly related to the inefficient use of these resources. Cost analysts ought to consider the following operational cost risk factors when pricing PFI/PPP facility management agreements:

- increase in unit cost of all FM cost centres;
- specification for cleaning;
- cost of specialised services, e.g., clinical waste disposal;
- availability of skilled labour;
- energy cost escalation;
- increase in demand for energy;
- increase in cleaning requiring;
- increase in security requirement;
- high staff turnover;
- increase in management costs.

Dealing with repair and maintenance cost risk

Maintenance cost is the expense of maintaining the building, to keep it in good repair and working condition. The expenditure on major maintenance and replacement is largely influenced by the constraints on the client's operating budget. The public sector often cannot afford the optimal maintenance of its building assets. The problems of incorporating uncertainty into operational cost and maintenance modelling have been addressed in several publications, which have applied probability theory to maintenance decision-making and cost modelling. For example, Wirahadikusumah *et al.* (1999) used a basic Markov chain model to aid decision-making in maintenance and rehabilitation of sewer systems. This model is unable to cope with changes in inflation and discount rates associated with the costs of decisions, therefore, the inclusion of sensitivity analysis would provide a better model. The Markovian property was also exploited by Scherer and Glagola (1994) who developed a Markov decision process for modelling the maintenance demands of infrastructure systems.

Similarly, Cesare *et al.* (1992) employed a comparable method whereby Markovian transition matrices are used to establish the current state of infrastructure systems so that future maintenance decisions can be assessed more effectively. Williams and Hirani (1997) described a model that determines the optimal, multi-level inspection-maintenance policy for a stochastically deteriorating, multi-state sub-system. The sub-system in this model is assumed to be a non-decreasing semi-Markov process. Similarly, Fung and Makis (1997) present a Markov renewal theory-based model for assessing deterioration and failure times in machinery components. Markovian models for bridge maintenance management were presented by Scherer and Glagola (1994), who identified the strength of Markov process in decision-making techniques. Other non-stochastic theories for dealing with maintenance are described in Chapter 11. As stated in Chapter 11, unfortunately, the majority of existing PFI/PPP facility management agreements have ignored the issues of basing maintenance regimes on condition surveys.

In PFI/PPP schemes, normally a life cycle fund is established. The fund is used for the major maintenance and replacement of building components that are going to require significant expenditure over the contract period. The existence of such reserved funds will go a long way to re-addressing the problem of lack of maintenance budgets that has led to the degeneration of a large stock of public building facilities and, certainly, it will benefit the clients of PFI projects, if well managed. However, the effectiveness of such a fund largely depends on the ability of the fund holder to resolve conflicts between capital and operational expenditures. Most PFI/PPP projects are only a few years into their operational periods. It is hard to say what impact this fund will have on the residual value of the asset.

Repair and maintenance cost risks are borne by whoever holds the life cycle fund (although minor maintenance aspects are, sometimes, dealt with under operation contracts, readers can find further details in Chapter 11). These risks are mainly due to:

- early failure of components and equipments;
- inadequate life cycle fund;
- errors in estimating the life cycle fund;
- increase in building occupancy level;
- unavailability of skills and expertise to carry out maintenance and replacement regimes;
- poor quality maintenance regime;
- unanticipated high demand on M & E equipment and systems;
- obsolescence of plant and equipment before the end of their anticipated service life;
- unexpected increase in planned maintenance;
- unexpected increase in routine maintenance;
- unexpected increase in life cycle replacement.

The project company will, through the concession agreement, assume responsibility for providing a specified service for the public sector, complying with a pre-set minimum service level. To achieve this, operating and maintenance risks are passed on from the project company to the operator(s) through operation and maintenance agreements, that are designed to ensure that the facilities are operated and maintained at least at the agreed minimum level.

Pricing the risk

Most authors in the construction industry sector tend to concentrate on the effectiveness of risk management strategies and to some extent ignore or underestimate the price of risk and its impact on whole life cost of building assets. It is imperative that cost analysts identify the costs of risks and determine the most effective strategies to minimise the impact of these costs on the whole life cost of building facilities. An important concept in risk management is risk pricing, which is often referred to as value at risk in the financial sector. Pricing risk is the estimation of the maximum loss that could be incurred by a particular aspect of the PFI/PPP deal (e.g., design, construction, operation, etc.). The price of risk is derived from all possible direct and indirect losses that may occur as consequence of risk impact. The price of risk associated with PFI/PPP schemes is complex, dynamic and continuous throughout the concession agreement. Hence, it is very difficult to price the risk of every stage of the project development with a great deal of certainty. In the majority of PFI/PPP cases, risk is priced as probable costs. In pricing risk, the cost analyst must carefully consider both scenarios of potential gain and loss from retaining or transferring risks. The rule of thumb is that the more risk that a stakeholder accepts, the higher the potential gains and losses. But, in PFI/PPP projects, risk analysts are mainly concerned with potential losses and their impact on service delivery to the client as well as impacts on the unitary payment and debt servicing. Based on this paradigm, the risk analyst ought to identify and price all potential risks throughout the life cycle of the concession agreement. The analyst must determine the expected risk price for all critical activities of the project at pre- and post-financial close. It is essential, here, that the stakeholder who is taking the risk must determine what is the maximum acceptable loss that he can sustain without affecting his business activities and liquidity. The potential loss could be direct or indirect, e.g., significant increases in insurance fees. This is important and it is in the best interest of the whole PFI/PPP stakeholders, that the risk bearer should be able to sustain any potential loss as well as remaining in liquidity (in operation). Stakeholders' liquidity is directly related to their capital adequacy to meet all accepted risk exposure either through insurance or through capital assets and current cash reserves.

The level of risk value is related to the strategy that might be adopted to manage the risk. Transferring risk through insurance policies and subcontracting will have the effect of reducing the cost of risk. Also adopting strategies that completely avoid risks will lead to a significant reduction in the risk costs.

In pricing risks, cost analysts have to take into consideration the balance between the cost of risk transfer and the cost of losses, if risk is retained. The decision-maker must look at all available options and evaluate their cost. The aim of this deliberation is to optimise the cost of risk transfer. In the quest to find the optimum risk cost, analysts must consider all possible scenarios including:

- minimum potential risk price;
- maximum potential risk price;
- expected risk price.

The risk prices should include both direct (e.g., insurance, payments, delays, etc.), and indirect prices (e.g., increase in insurance premiums, loss of market).

Having examined all potential risk prices, the analysts must now decide on the optimum level of risk transfer. The optimum level of risk transfer is dependent on a trade-off between the cost of risk and the cost of losses that might be sustained by a particular PFI/PPP stakeholder, if risk is retained. This process is very complex and depends on many variables. In any case the guidelines from the UK Government advice to procuring authorities is to only transfer a risk when it can obtain value for money by such a transfer. Analysts might consider using sensitivity analysis or even better simulation methods to derive the optimum risk price and the level of risk to be transferred.

One of the most difficult tasks in pricing risks is the availability of reliable cost information on risk impacts or prices. Having said this, risk can be priced based on both hard data and on the basis of subjective judgement from past experience. The pricing can be done based on the following strategies:

- *Single point estimate*: the approach adopted here is to identify key risks and then estimate the price of each risk and risk category in their own right. The probability of the occurrence of risks is also estimated. The price of risk can be estimated either as a percentage excess cost of the base cost or simply as an estimated loss if such risk occurs. The analyst then multiplies the risk impact (price) by the probability of the occurrence to determine the anticipated risk estimates. These estimates are then added to compute the risk price for each risk category. The cumulative risk price of all risk categories will form the total project risk price. This price is then used to adjust the project NPV to risk. It is important here that the risk analyst must ensure that the assumptions and the reasoning behind the risk estimates are fully understood and properly documented for traceability purposes. An example of single point estimate is shown in Table 12.1.
- *Three point estimates*: this method requires the risk analyst to define or

Table 12.1 An example of risk single point estimates

Risk description	*Occurrence year*	*Expected value £'000*	*Observation notes*
Design			
Failure to design to brief	1	12	Loss in performance leads to implementation delays
Continuing development to design	1 to 2	10	Delays in design lead to additional construction time and cost
Change in design by operator	2 to 3	13	Lead to service delay and additional costs
Construction			
Time overrun	3	45	Delays in construction results in additional cost and delay start of services
Unforeseen site conditions	1	15	Expect additional costs and delay in construction
Delay in gaining access to site	1 to 3	20	Delays in construction and delay start of services
Contractor/sub-contractor dispute	1 to 3	30	Disputes lead to additional cost and delays for both construction and start of services
Operating Cost Risks			
Incorrect estimated costs	4 to 30	18	Soft FM services costs may increase or decrease following market testing
Changes in taxation	4 to 30	10	Any legislative or regulatory change could lead to additional costs
Non-performance of services	4 to 30	125	Creates additional costs in the provision of services
Availability and Performance			
Latent defects in new build	4 to 6	65	Additional cost at first year of planned service operation
Performance of sub-contractors	4 to 30	37	Affects the operation and maintenance costs
Availability of facilities	4 to 30	70	Affects service delivery and payment

estimate a lower and upper limit to a most likely risk price for every risk variable. The degree of variability, i.e., range between the risk prices, is indicative of the level of certainty about the cost of a particular risk factor. Risk analysts must consider all possibilities, including extreme conditions, and should aim to define realistic ranges. The analyst ought to consider estimating the probability of risk occurrence as three point estimates as well. Having developed the assumption and estimated risk cost ranges, preferably through several workshops, the analyst may choose to use the average or weighted average of the three point estimates to calculate the price of risk. The estimated risk prices are normally discounted to NPV before any further modelling or analysis is carried out on them. The analyst has several options on how to arrive at the most probable risk costs. First, he can use a simple method of multiplication and addition to arrive at the overall risk price for each risk factor. Second, he can use triangular distributions for the risks and simulation models, e.g., Monte Carlo, to produce probability values of risk for the project at different percentile values of risk.

- *Mathematical functions*: instead of the above simplified methods, some risk analysts may prefer to use more sophisticated mathematical functions to model the variation in the risk costs. These methods are rarely used in the construction industry sector.

Whatever the method adopted by the risk analyst, the pricing of risk is a progressive and iterative process that takes several workshops to complete, involving experts from all aspects of design, construction and operation of building assets. Tables 12.1–4 show examples of risk identification, pricing and modelling based on the above scenarios. Note that in all the tables, estimates of risk costs are discounted to the NPV value in year zero to drive the estimate of the NPV of each risk.

- *Table 12.1*: illustrates the use of single point estimate for pricing the risk of a PFI/PPP deal. Simply here, a risk value is estimated for each risk factor and then added to arrive at the total NPV value of the risk.
- *Table 12.2*: shows how risk costs can be estimated probabilistically. This is a very simple way of deriving the impact of risk based on their magnitude of impact and possibility of occurrence. Probably this method is most widely used in PFI/PPP schemes risk analysis. The method shown in this table can be modified to include three point estimates of each risk value and their associated probability of occurrence.
- *Table 12.3*: demonstrates the use of probability and the three point estimate method for cost risk factors. Notice in this example the average estimate risk is used to derive the expected risk costs. In many cases it is sufficient to produce this type of estimate of risk cost and to discount this to produce a range of the NPV of the risk. These range values can then be

Table 12.2 An example of probabilistic risk pricing

Risk description	*Occurrence year*	*Probable value £'000 1*	*Probability occurrence % 2*	*Expected value of risk £'000 3 = 1*2*
Design				
Failure to design to brief	1	12	0.50	6
Continuing development to design	1 to 2	10	0.30	3
Change in design by operator	2 to 3	13	0.20	3
Construction				
Time overrun	3	45	0.30	14
Unforeseen site conditions	1	15	0.75	11
Delay in gaining access to site	1 to 3	20	0.40	8
Contractor/sub-contractor dispute	1 to 3	30	0.20	6
Operating Cost Risks				
Incorrect estimated costs	4 to 30	18	0.35	6
Changes in taxation	4 to 30	10	0.70	7
Non-performance of services	4 to 30	125	0.30	38
Availability and Performance				
Latent defects in new build	4 to 6	65	0.60	39
Performance of sub-contractors	4 to 30	37	0.15	6
Availability of facilities	4 to 30	70	0.45	32

Note: Probable risk values are already time-adjusted.

used in sensitivity analysis or alternatively as the input to the risk simulation model for the project.

- *Table 12.4*: presents an example of how to combine three point estimates (triangular distribution) and simulation models to derive risk prices. The three point estimates of the risks are set out against the project timescale. Output from such models could include risk value distributions, standard deviation, median and maximum percentile values and any other statistical measures.

The above are just a few scenarios on how risk costs can be derived. The selection of a particular method largely depends on project complexity and the skills of the risk analyst.

Summary

It is a fundamental principle of PFI/PPP procurement that risk assessment is made and risk is transferred to parties that are able to bear the financial and managerial consequences, thus meeting the criteria of value for money. Risk is a

Table 12.3 The use of probability and three point estimates in risk pricing

Risk description	*Occurrence year*	*Probable risk value £'000*	*£'000*	*£'000*	*Average value £'000*	*Probability occurrence %*	*Expected value of risk £'000*
		MN	*ML*	*WC*			
		1	*2*	*3*	*4 = (1 + 2 + 3)/3*	*5*	*6 = 4*5*
Design							
Failure to design to brief	1	10	12	20	14	0.50	7
Continuing development to design	1 to 2	5	10	15	10	0.30	3
Change in design by operator	2 to 3	10	13	25	16	0.20	3
Construction							
Time overrun	3	30	45	80	52	0.30	16
Unforeseen site conditions	1	10	15	30	18	0.75	14
Delay in gaining access to site	1 to 3	15	20	50	28	0.40	11
Contractor/sub-contractor dispute	1 to 3	30	30		20	0.20	4
Operating Cost Risks							
Incorrect estimated costs	4 to 30	15	18	40	24	0.35	9
Changes in taxation	4 to 30	5	10	30	15	0.70	11
Non-performance of services	4 to 30	100	125	200	142	0.30	43
Availability and Performance							
Latent defects in new build	4 to 6	50	65	100	72	0.60	43
Performance of sub-contractors	4 to 30	30	37	60	42	0.15	6
Availability of facilities	4 to 30	40	70	120	77	0.45	35

Notes: Probable risk values are already time-adjusted.
MN = minimum; ML = most likely;
WC = worst case, these values are used to construct triangular distributions.

Table 12.4 The use of simulation models to price risk

Risk description	*year 1*			*year 2*			*year 3*			*year 4*			*year 5*			*year n*	*year 30*			*year 30*		
	cost of risk			*cost of risk*			*cost of risk*			*cost of risk*			*cost of risk*				*cost of risk*			*total NPV*		
	MN £'000	*ML £'000*	*WC £'000*	*MN £'000*	*ML £'000*	*WC £'000*	*MN £'000*	*ML £'000*	*WC £'000*	*MN £'000*	*ML £'000*	*WC £'000*	*MN £'000*	*ML £'000*	*WC £'000*		*MN £'000*	*ML £'000*	*WC £'000*	*MN £'000*	*ML*	*WC*
Design																						
Failure to design to brief	10	12	20																	10	12	20
Continuing development to design	5	10	15	5	10	15														10	20	30
Change in design by operator	10	13	20	5	7	10	1	5	6											16	25	36
Construction																						
Time overrun							30	45	80											30	45	80
Unforeseen site conditions	10	15	30																	10	15	30
Delay in gaining access to site	15	20	50	2	5	10														17	25	60
Contractor/sub-contractor dispute							30	30	50													
Operating Cost Risks																						
Incorrect estimated costs													1	2	3					15	18	40
Changes in taxation																	1	3	5	5	10	30
Non-performance of services													1	2	3		1	3	4	5	15	40
Availability and Performance																						
Latent defects in new build													5	6	7		5	10	20	50	65	100
Performance of sub-contractors													1	2	3		1	3	4	30	37	60
Availability of facilities													1	2	3		1	3	4	40	70	120

Notes probable risk values are already time adjusted
MN = minimum, ML = most likely, WC = worst case, these values are used to construct triangular distributions.
Data in this table are designed as an input to a risk model which uses Monte Carlo simulation.

key aspect of PFI/PPP projects and provides the basis for much of the contract and for demonstrating value for money. Risk costing is also fundamental in both the management and economical evaluation of PFI/PPP projects. The use of risk assessment techniques in the pricing of PFI/PPP deals vary from project to project. What is appropriate depends on the relative size of the project, availability of data and resources, computational aids and skills, and cost analyst understanding of the technique being applied. The author always advocates the use of a combination of techniques to estimate risk costs. This chapter has demonstrated that the risks can have a significant impact on the value of analysis aspects of the project. There is also evidence to show that risk costs attributed to public clients are significantly higher than those assigned to the private sector. Operational, maintenance and life cycle replacement risks are controlled and managed according to the specific project agreement clauses. The chapter concluded by presenting several options for pricing risks in PFI/PPP deals. The author believes that the pricing of risk is a progressive and iterative process that needs to be cautiously considered by risk analysts.

Chapter 13

Cost risk allocation in PFI/PPP projects

Introduction

Risk quantification, pricing and allocation are the foundation on which PFI/PPP procurement is based. In PFI/PPP procurement, decisions are based on concepts of risk transfer and value for money. The UK Government stated that PFI/PPP will deliver value for money over the life of the projects because the private sector assumes some of the financial risks (and costs) that the public sector would otherwise carry (HM Treasury 2004b). This chapter explores the cost risk allocation mechanisms in PFI/PPP projects. First, it sets out the principles of risk allocation. Second, it describes the process involved in different risks' allocation and transfer. Third, it explains and presents the available mechanisms for dispute resolution in such a complex procurement system.

Principles of risk allocation

The risks associated with design, construction, finance, maintenance and operation of the scheme over the life of the project must be identified, quantified, priced and transferred to the appropriate party. Risk allocation has been identified as one of the most important elements of PFI/PPP procurement. Conventional risk allocation practices advocate that risks ought to be transferred to the party who can best control them. However, this party, in practice, may not be sufficiently financially robust to absorb the cost of the allocated risks.

Risk transfer to the private sector is important in the demonstration of value for money and determining the balance sheet treatment in PFI/PPP schemes. Froud (2003) states that the notion of risk is central to the evaluation of PFI/PPP projects for affordability:

> Risk is central to understanding and analyzing PFI. The existence of risk in the sense of possibility of things going wrong is paradoxically a positive force providing the motivation for and possibility of new forms of finance for public infrastructure.

Froud assumes that through drawing up contracts that identify, allocate and manage risk, the demonstration of value for money is made possible. Akintola *et al.* (1998) stated that PFI/PPP projects require an appropriate transfer of risk to the private sector, both through the design, planning and construction phases and, in operation, through a combination of payment mechanisms and specific contract conditions.

The principles of risk allocation have been the subject of debate in the industry for many years. Various authors have dealt extensively with risk allocation principles. For example, Barnes (1983) has developed an algorithm for risk allocation. The method is based on the following iterative steps:

1 Develop a list of the unrelated risks that have to be carried by one or other of the parties.
2 Identify the risks that are outside the contractor's control. Allocate these to the client.
3 Rank the list in order of magnitude (measured as the standard deviation of cost uncertainty).
4 Add the risk (taking the square root of the sum of the squares), working from the largest first and noting the cumulative total. Stop when the cumulative total levels out.
5 If the cumulative total exceeds a tolerable threshold (e.g., 10 per cent of the estimate base cost), consider what steps could be taken to manage the risk. Go back to step 3 and continue.
6 If the cumulative total is less than the threshold, allocate the remaining large risks and all the small risks to the contractor.

Barnes stated that if the above process for risk allocation is carried out properly, it will result in the risk that is borne by the contractor being enough to sustain his incentive, but not large enough to be detrimental to his business activities.

Lewendon (2006) listed the fundamental principles of risk allocation as follows:

- Risks should be allocated to the party best able to control them prior to their occurrence.
- Risk allocation should encourage good management by the party who carries the risk.
- Motivation is provided by the financial consequence.
- Risks should not be allocated to a party who is unable to sustain their consequences. Risks which are outside the contractor's control should usually be allocated to the employer.
- Where risk allocation is split, the split should reflect each party's ability to influence the likelihood of occurrence and effect.
- The contract should not be cluttered up with risks of small likelihood and impact.

The NHS (1995) has produced a list of 20 risk questions for assessing if a private partner could protect the NHS against these risks. These questions are used to derive risk transfer strategies. These are:

1 Construction costs overrunning?
2 Losses through completion delay?
3 Costs of latent defects?
4 Losses through unavailability of any aspect of facilities?
5 Escalating maintenance and repair costs?
6 Failure to meet energy-efficiency targets?
7 Failure to meet facilities management cost targets?
8 Income generation schemes failing to meet net income targets?
9 Costs of new investment to maintain performance of income generation schemes?
10 Quality standards of facilities failing to meet pre-set performance targets?
11 Quality standards failing to keep up with new levels being achieved by competitors?
12 Losses through shortages of key inputs?
13 Problems through facilities failing to keep pace with new technology?
14 Problems through design of facilities hindering effectiveness?
15 Losses through design of facilities proving inefficient in use?
16 Escalation of general operating costs?
17 Losses through costs exceeding competitors' prices?
18 Losses through facilities proving too big or too small for needs?
19 Costs of adaptation for alternative use?
20 Lower than expected residual or sale value?

The process of risk allocation

The aim of whole life risk allocation in PFI/PPP projects is to determine whether public projects should be funded by public or private finance, or both. The process of risk allocation plays a vital role in this process. The process of risk allocation is shown in Figure 13.1. The process consists of the following iterative steps.

Risk register

It is important that the risk register should be used as a tracking device to manage risk throughout the life of a concession. The risk register is normally populated with information from the early stages of the PFI/PPP procurement process. It is a requirement that all procuring authorities must provide a risk register at the outline business case and for all subsequent stages of the procurement. The risk register is an interface part of the appraisal documentations. It is expected that risk registers are updated as more information becomes

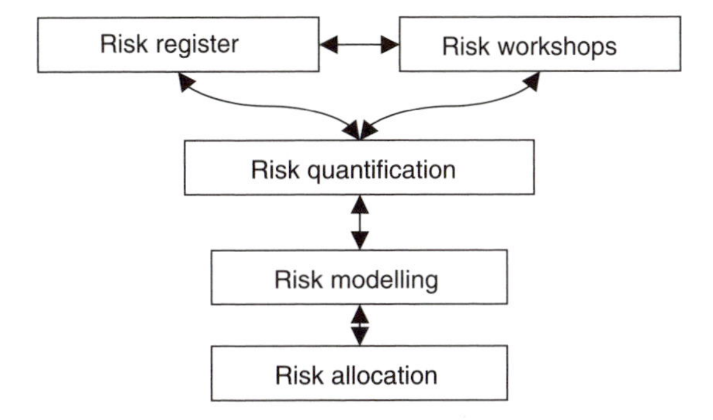

Figure 13.1 The process of risk allocation.

available during the procurement process. Risk registers must not be used only as a medium for recording risks. The register must be treated as a live document for tracking how registered risks are resolved and managed. To ensure the effectiveness of the tracking process, each risk should be assigned to an owner who will be responsible for making sure risks are managed as stated in the register.

In PFI/PPP schemes, the risk register starts from design and construction and then moves into operational risks. Risk registers enable the riskiness of the PFI/PPP projects to be early identified and tracked over a period of the concession agreement time to show how effectively the risks have been priced and allocated.

The purposes cited for risk registers include:

- assurance control;
- repository knowledge;
- risk allocation;
- risk transfer;
- medium for interface between project phases;
- risk tracking;
- risk management;
- risk reduction.

Standard risk registers have been issued by the UK Government to be used in PFI/PPP contracts. For example, the purpose of a risk register is explained in the HM Treasury Green Book (2004b) as follows:

> A risk register lists all the identified risks and the results of their analysis and evaluation. Information on the status of the risk is also included. The risk register should be continuously updated and reviewed throughout the course of a project.

The OGC's effective partnering guide (2002b) says that:

> A shared risk register ensures complete understanding for both parties about risks to implementation and ongoing service delivery, and enables a joint approach to managing risks. Clarity of who is responsible for, and manages, which risks is also essential.

According to the HMT *Green Book*, the risk register ought to include the following data:

- risk number;
- risk type;
- author (who raised it);
- date identified;
- date last updated;
- description;
- likelihood;
- interdependencies with other sources of risk;
- expected impact;
- bearer of risk;
- countermeasures;
- risk status and risk action status.

Risk workshops

In parallel with the risk register creation, risk workshops are held to assist in the process of risk identification, quantification and allocation. This is important, especially if risk analysis is based on professional judgements in a multidisciplinary environment. Qualitative approaches are useful in developing ranges for risk prices rather than single risk costs. However, the author advocates the use of probabilistic risk analysis to support qualitative judgements.

The format and duration of risk workshops vary depending on the complexity of the project under analysis. But, in general, the format tends to be based on brainstorming and Delphi techniques. Usually workshops are facilitated by expert consultants to direct and stimulate discussion among the participants. Checklists and prompt cards are used to ensure all important risk aspects are discussed. The number of participants and their expertise vary from project to project as well as from stage to stage throughout the project life cycle. Complex projects will require several workshops to create the project risk register. At the end of risk workshops, participants are expected to identify, score and rank project risks. The ranking and scoring processes capture the collective knowledge and experience of the participants. Hopefully, the outcome will form a consensus on which project risk is quantified and allocated. The information generated from this exercise is used to create a risk register. The risk analyst must appreciate that the evaluation of risk is a continuous process. Hence, further workshops may be required at different project milestones, for example, at the financial close and project agreement negotiation.

Risk quantification

The process of risk pricing and risk quantification methods are discussed in detail in Chapter 12. It is widely accepted that risk quantification is defined as the magnitude and time frame of each risk event. This process is set out to quantify the impact of the risks identified in the risk register. Probability impact matrices are used extensively for these propose. Hence, the purpose of risk quantification is to appreciate and estimate the likelihood of occurrence and the potential impacts on the PFI/PPP project cost outturn. For example, RAMP (2006) guidelines stipulate that probability and impact of risks should be scored on a calibrated scale and the risk is quantified by multiplying the probability by the anticipated impact. According to the OGC (2003a), both qualitative and quantitative methods should be used in risk quantification. Qualitative quantification methods are used to do the following:

- assess time of risk occurrence;
- identify elements of the project that could be affected;
- identify cause–effect relationships;
- determine the likelihood of risk occurrence;
- determine how the risk will affect the project.

Quantitative assessment methods are used to quantify the impact of risk occurring in terms of impact on the following:

- the base cost and project outturn cost;
- the base estimate of the completion date for the project;
- the project performance not meeting the user requirements;
- the health and safety requirements;
- sustainability and design quality implications.

Gadd (2001) recommended that the following issues be considered in risk quantification:

- Make use of all available data and data sources.
- Make use of appropriate experts.
- Ensure that the assumptions and the reasoning behind the risk estimates are fully understood and properly recorded.
- Estimation should be conducted on the basis of a range estimate.
- The estimation of risk is a progressive and iterative process that takes more than a single workshop to complete.

- Do not expend valuable resources chasing unnecessary detail and accuracy that will not have a material effect on the investment appraisal.

Risk modelling

To assist in the process of risk modelling, there are several computer packages which allow the risk identified in the risk register to be modelled and the likely risk costs to be determined. Several spreadsheet-based packages exist, including @Risk Crystal Ball, and Predict, which enable Monte Carlo simulation to take place in conjunction with the stages of the PFI/PPP project's life cycle. These packages can easily be used to determine the spread in risk variables throughout the concession agreement. This will assist in risk pricing and allocation.

Risk allocation

The principles of risk allocation are described on p. 276. According to the OGC (2002), '*risk allocation is about deciding who is best placed to manage a specific risk*'. The process here deals with the options available for risk allocation. In PFI/PPP procurement the process is based on the paradigm that risks are allocated to either the public or private sector depending on the type of risk and if the risk allocation solution delivers the most cost-effective option. The risk register normally sets out the risks that need to be allocated or transferred to the private sector. The register, and the contract for that matter, should be very clear as to which risks have been retained and which have been allocated to the project company. As stated above, the final allocation of risks may not be established until the end of the negotiation period, including due diligence, and financial close.

Risk classification

There are several ways of classifying risks in PFI/PPP projects. The most widely used risk categorisation methods are based on classifying risks according to their source, for example, cost risks, planning risks, etc. It is also possible to classify risks according to the life cycle of building assets development and operation. In some projects, risks are categorised at a strategic level as political, economic/commercial, environmental and social risks. Others take a more technical view by using categories such as controllable and uncontrollable risks.

Several categorisations of risks in PFI/PPP type projects have been published, for example by the HM Treasury (1997b) and the State of Victoria (2001). Bing *et al.* (2004) clustered PFI/PPP project risks into macro-level risks (risks external to the project), meso-level risks (risks that occur within the project organisation) and micro-level risks (risks that originate from stakeholders' relationships). This type of classification is unhelpful and has little value in risk assessment and pricing. Grimsey and Lewis (2002) cited nine categories of high-level risk that are common for all infrastructure projects. These categories include the following risks:

- technical
- construction
- operating
- revenue
- financial
- *force majeure*
- regulatory and political
- environmental
- project default
- asset ownership.

The National Audit Office (1999a) listed the main categories of risk-involved PFI/PPP projects as:

- design and construction risks;
- commissioning and operating risks;
- demand risks;
- residual value risk;
- technology/obsolescence risks;
- regulation risks;
- project financing risks;
- risk of contractor default;
- political/business risk.

Gadd (2001) emphasised that the above risks are important in PFI/PPP projects for the following reasons:

- They affect the ability of those funding the project to recover their investment and hence will be closely scrutinised by the funders.
- They affect to some extent the incentive on the contractor to perform and to maintain the quality of the service and service assets through the life of the project, in particular demand risk and under-performance or non-availability.
- They affect the balance sheet treatment of the project.

Not all of the above risks are relevant to every project. Careful consideration needs to be given to whether the effects of these risks are in terms of their possible occurrence and impact on the project. Therefore understanding which risks are possible enables effort to be focused on appropriate classes of risks. It is a requirement that every project that is considered for PFI/PPP finance should be assessed for risks. A clear understanding of the way in which these risks affect the ability of the private sector to deliver the service is paramount to the delivery of value for money.

The NAO and NHS risk classification is adopted by this chapter to explain the process of risk allocation in PFI/PPP projects. These risks normally form the primary category (headings) of the PFI/PPP risk allocation matrix. Gadd (2001) explained the benefits of the risk allocation matrix as follows:

> Establishing a risk allocation matrix is a fundamental requirement for an effective risk allocation and management process The matrix provides a structure that ensures all areas of risk are considered and that risk is addressed consistently across all supplier negotiation.

Normally, risk allocation matrices are drafted prior to the issue of the ITN documentations. The risk matrix plays a pivotal role in the procurement process. All contract negotiations are based on the risk trade-off between risk allocation, pricing and incentives. The key risks that might be addressed under each of the matrix headings are discussed in turn in more detail in the following sections.

Allocation of design and construction risks

Design and construction risks are considered together due to the fact that PFI/PPP projects are contracted under the design and build procurement system.

Design and construction risks are defined by the State of Victoria (2001) as:

> The risk that the design, construction or commissioning of the facility or certain elements of each of these processes, are carried out or not carried out in a way which results in adverse cost and/or service delivery consequences. The consequences if the risk materialises may include delays and/or cost increases in the design, construction and commissioning phases, or design or construction flaws which may render the infrastructure inadequate for effective service delivery, either immediately or over time.

From this definition, it is clear that design risk stems from the risk that the design is unable to deliver the services at the required performance or quality standards. Construction risks are related to contractors' lack of performance, which may result in adverse cost and service deliver consequences, either immediately or over the concession period. Design and construction risks can be in the form of cost escalations, time delays and construction defects in the facility provided. The individual risks occurring under this heading are listed by the NAO (1999a) as:

- Surveys and investigations fail to identify problems.
- Construction lasts longer than expected.

- Construction costs are higher than excepted.
- Facilities are not provided to the required specification.
- Alternative service provision is required during the design completion.

All design and construction cost risks are expected to be borne by the private sector partner. NAO (1999a) stated that: 'The private sector contractor should meet any cost increase arising from poor estimating, delays and failure to meet the client's requirements, and should not be paid until the start of service delivery.'

It is imperative to realise here that as a result of this high risk, the price of the design and build contract in a PFI/PPP deal is likely to be higher than the market price for procuring the same projects independently. There are several issues that need to taken into consideration in allocating the risk under this heading. According to the State of Victoria (2001), this includes:

- The procurer must not share design and construction risks.
- The procurer bears the risk of any design or construction changes and all other interferences in the design and construction process initiated by him.
- If risks are transferred to a third party, the project company retains the primary liability for the particular risk in question.
- The private sector bears the cost consequences of delays in design, construction and any failure to meet the agreed specifications.
- Large changes in design and specification requirements could lead to the client taking over the responsibility for the design and may make the allocation of design risk to the private sector ineffective.
- It is the responsibility of the private sector to ensure the design/construction is capable of delivering the specified service outputs.
- A PFI/PPP building asset must be fit for purpose, i.e., to be fit for use as a venture to meet functional and service output requirements.

Issues such as design specification, construction phasing/milestones, commissioning deadlines, delays, delay events, deadlines, time extensions and construction defects are dealt with under the project agreement. Chapters 10 and 11 provide further details on risks from delays and delay events and how they are allocated and managed. All PFI/PPP project agreements are designed to allocate obligations and responsibilities to the parties involved. Table 13.1 shows which risks are transferred to the private sector and which are retained by the public authority. As can be seen from Table 13.1, all design and construction risks are allocated to the private sector with the exception of the 'Delay Events, *force majeure*, Termination due to *force majeure*' risk factors which are shared, and legislative/regulatory change (specific to the project) and compensation events risks are taken by the procurer.

Table 13.1 NHS LIFT Risk Register

No.	*Risk Heading*	*Definition*	*Allocation of Risk*		
			Tenant	*LIFTCo*	*Shared*
1. Design Risks					
1.1	Failure to design to brief	Failure to translate the requirements of the Tenant into the design.		✔	
1.2	Continuing development of design	The detail of the design should be developed within an agreed framework and timetable. A failure to do so may lead to additional design and construction costs.		✔	
1.3	Change in requirements of the Tenant	The Tenant may require changes to the design, leading to additional design and construction costs.	✔		
1.4	Change in design required by LIFTCo	This is the risk that LIFTCo will require changes to the design, leading to additional design costs.		✔	
1.5	Change in design required due to external influences specific to the provision of health and social care services.	There is a risk that the designs will need to change due to legislative or regulatory changes specific to the provision of health and social care services.			✔
1.6	Failure to build to design	Misinterpretation of design or failure to build to specification during construction may lead to additional design and construction costs.		✔	
2. Construction and Development Risks					
2.1	Incorrect time estimate	The time taken to complete the construction phase may be different from the estimated time.		✔	
2.2	Unforeseen ground/site conditions	Unforeseen ground/site conditions may lead to variations in the estimated cost.		✔	

2.3	Delay in gaining access to the site, 'Delay Events'	A delay in gaining access to the site may delay or impede the performance of the contract and cause additional expenses.	✔		
2.4	Responsibility for maintaining on site security	Theft and/or damage to equipment and materials may lead to unforeseen costs in terms of replacing damaged items, and delay.		✔	
2.5	Responsibility for maintaining onsite safety	The Construction, Design and Management (CDM) regulations must be complied with.		✔	
2.6	Third party claims	This risk refers to the cost associated with third party claims due to loss of amenity and ground subsidence on adjacent properties.		✔	
2.7	Compensation events	An event of this kind may delay or impede the performance of the contract and cause additional expense.	✔		
2.8	*Force majeure*	In the event of *force majeure*, additional costs will be incurred. Facilities may also be unavailable.			✔
2.9	Termination due to *force majeure*	There is a risk that an event of *force majeure* will mean the parties are no longer able to perform the contract.			✔
2.10	Legislative/regulatory changes specifically relating to the provision of health and social care	A change in legislation/ regulations not specific to the provision of health and social care services leading to a change in the requirements and variations in costs.			✔
2.11	Legislative/regulatory change not specific to the provision of health and social care services	A change in legislation/ regulations not specific to the provision of health and social care services leading to a change in the requirements and variations in costs.		✔	
2.12	Changes in taxation	Changes in taxation may affect the cost of the project.		✔	
2.13	Changes in the rate of VAT	Changes in the rate of VAT may increase the costs of the project.		✔	

(Continued overleaf)

Table 13.1 Continued

No.	*Risk Heading*	*Definition*	*Allocation of Risk*		
			Tenant	*LIFTCo*	*Shared*
2.14	Other changes in VAT	Changes in VAT legislation other than changes in the rate of VAT payable.		✔	
2.15	Contractor default	In the case of contractor default, additional costs may be incurred in appointing a replacement, and may cause a delay.		✔	
2.16	Poor project management	There is a risk that poor project management will lead to additional costs. For example, if subcontractors are not well co-ordinated, one subcontractor could be delayed because the work of another is incomplete.		✔	
2.17	Contractor/sub-contractor industrial action	Industrial action may cause the construction to be delayed, as well as incurring additional management costs.		✔	
2.18	Protestor action	Protestor action against the development may incur additional costs, such as security costs.			✔
2.19	Incorrect time and cost estimates for commissioning new buildings	The estimated cost of commissioning new buildings may be incorrect, there may also be delays leading to further costs.		✔	
3. Availability and Performance Risks					
3.1	Latent defects in new buildings	Latent defects to the structure of the new builds, which require repair, may become patent.		✔	
3.2	Change in specification initiated by procuring entity	There is a chance that, during the operating phase of the project, the procuring entity of the services will require changes to the specification.	✔		

3.3	Performance of subcontractors	Poor management of subcontractors can lead to poor co-ordination and under-performance by the contractors. This may create additional costs in the provision of services.	✔	
3.4	Default by contractor or subcontractor	In the case of default by a contractor or subcontractor, there may be a need to make emergency provision. There may also be additional costs involved in finding a replacement.	✔	
3.5	Industrial action	Industrial action by the staff involved in providing facilities services would lead to higher costs and/or performance failures.	✔	
3.6	Failure to meet performance standards	There is a risk that facilities management (FM) will not provide the required quality of services. This may be costly to correct, and LIFTCo may incur financial penalties.	✔	
3.7	Availability of new facilities	There is a risk that some or all of the facility will not be available for the use to which it is intended. There may be costs involved in making the facility available.	✔	
3.8	Relief Events	An event of this kind may delay or impede the performance of the contract and cause additional expense.		✔
3.9	*Force majeure*	In the event of *force majeure*, additional costs will be incurred. Facilities may also be unavailable.		✔
3.10	Termination due to *force majeure*	There is a risk that an event of *force majeure* will mean the parties are no longer able to perform the contact.		✔
4. Operating Cost Risks				
4.1	Incorrect estimated cost of providing specific services under the contract	The cost of providing these services may be different to those expected, because of unexpected changes in the cost of equipment, labour, utilities and other supplies.	✔	

(Continued overleaf)

Table 13.1 Continued

No.	*Risk Heading*	*Definition*	*Allocation of Risk*		
			Tenant	*LIFTCo*	*Shared*
4.2	Legislative/regulatory change having capital cost consequences specifically relating to the provision of health and social care	Changes specifically relating to the provision of health and social care in legislation/ regulations may lead to additional construction costs, and higher building, maintenance, equipment or labour costs.			✔
4.3	Legislative/regulatory change not specific to the provision of health and social care services	Changes not specific to the provision of health and social care services in legislation/ regulations may lead to additional construction costs, and higher maintenance, equipment or labour costs.		✔	
4.4	Changes in taxation	The scope and level of taxation will effect the cost of providing services.		✔	
4.5	Changes in VAT	This may increase the cost of the provision of services.		✔	
4.6	Incorrect estimated cost of providing clinical services	The cost of providing clinical services may be different to the expected. These costs include: staff, recruitment, training, equipment and supplies.	✔		
4.7	Incorrect estimated cost of maintenance	The cost of building and engineering maintenance may be different to the expected costs.		✔	
4.8	Incorrect estimated cost of energy used	Failure to meet energy-efficiency targets or to control energy costs.			✔
5. Variability of Revenue Risk					
5.1	Non-performance of services	Payment will only be made by the Tenant for services received.		✔	
5.2	Poor performance of services	The operator will incur deductions from the performance payment for the poor performance of services.		✔	

5.3	Changes in the size of the allocation of resources for the provision of health and social care	There is a risk that the resources allocated to the area are reduced or increased. If such changes do occur, there may be a need to re-scale the provision of services.	✔		
5.4	Changes in the volume of demand for customer services	There is a risk that the volume of demand for health and social care will change, because of changes in the size of the catchment area. This may occur because there is, for example: an unexpected increase in the size of the population, leading to an increase in demand; or the streamlining of provision may lead to an increase/reduction in demand.	✔		
5.5	Unexpected changes in technology	Unexpected changes in technology may lead to a need to re-scale or reconfigure the provision of services.	✔		
5.6	Unexpected changes in the demographics of the catchment area	Unexpected changes in the demographics of the catchment area may lead to a reconfiguration or re-scaling of the provision of services.	✔		
5.7	Unexpected sudden increase in demand	There is a risk of large unexpected increases in demand (e.g. due to a major incident).			✔
5.8	Estimated income from income generating schemes is incorrect	There is a risk that income generating schemes, such as car parking, generate less or more income than expected.		✔	
6. Termination Risks					
6.1	Termination due to default by the procuring entity	The risk that the procuring entity defaults leading to contract termination and compensation for the private sector.	✔		
6.2	Default by LIFTCo leading to step-in by financiers	The risk that LIFTCo or individual Contractors default and financier's step-in leading to higher costs than agreed in the contract.		✔	

(Continued overleaf)

Table 13.1 Continued

No.	*Risk Heading*	*Definition*	*Allocation of Risk*		
			Tenant	*LIFTCo*	*Shared*
6.3	Termination due to default by LIFTCo	The risk that LIFTCo defaults and step-in rights are exercised by financiers but that they are unsuccessful, leading to contract termination.		✔	
7. Technology and Obsolescence Risks					
7.1	Technological change/building obsolescence	Buildings, plant and equipment may become obsolete during the contract.		✔	
7.2	Technological change/building obsolescence	Technical changes may cause the Tenant to revise its output specification.	✔		
8. Control Risks					
8.1	Control of health and social care services	The Tenant retains control of health and social care services, which means that it retains significant control of the nature of the services provided by LIFTCo.	✔		
8.2	Control of services provided under the LIFTCo contract	LIFTCo should retain control of these subject to 8.1 above.		✔	
9. Residual Value Risks					
9	Procuring entity no longer requires buildings at end of contract	The risk that the procuring entity will wish to vacate the building at the end of the contract period, and that LIFTCo may be faced with decommissioning costs.		✔	
10. Other Project Risks					
10.1	Incorrect cost estimates for planning approval	Estimated cost of receiving detailed planning permission is incorrect, including the cost of satisfying unforeseen planning requirements.		✔	
10.2	Delayed planning approval	A delay in receiving planning permission may have broader cost implications for the project, as well as the loss of potential savings.			✔

Source: Reproduced with permission of 4Ps.

Allocation of commissioning and operational risks

The operational risk is attributed to the events that may affect the operation and maintenance of the building asset in a way that prevents the service provider from delivering the contracted services according to the project agreement specification and within the agreed costs. These risks are normally clustered under availability and performance risk. According to NAO (1999a) and the State of Victoria (2001), commissioning and operational risks that are found in a PFI/PPP project include:

- Contractor fails to meet performance standards for service delivery.
- Contractor fails to make assets available for use.
- Operating costs are more than expected.
- Operating costs are less than expected.
- Assets underpinning services delivered are not properly maintained.
- Unsuitable design.
- Higher maintenance costs.
- Occupational health and safety issues.
- Inherent defects.
- Technical obsolescence.
- Incorrect estimated cost of maintenance.
- Changes in taxation and VAT.

All service and maintenance cost risks and most operating risks are accepted by the private sector. However, the inflation risk, i.e., the risk that operating costs will increase, is shared through an annual indexation of the contract price once the provider starts to deliver the contracted services. Chapter 11 provides further details on benchmarking and market testing for service provision. Other issues that need to be taken into consideration in operational risks allocation, according to the State of Victoria (2001), include:

- If intervention by the client leads to operational cost increases, then the client ought to share such increased costs.
- The client bears the operating risk of the core service, e.g., clinical, teaching.
- The client bears the risk if the provision of the core service will impact on the private sector's ability to deliver the contracted services.

Table 13.1 shows which commissioning and operational risks are transferred to the private sector and which are retained by the client. As can be seen, from commissioning risks only, relief events (see Chapter 11 for further explanation) and 'change in specification initiated by procuring entity' risks are borne by the client.

Allocation of project finance risk

Project finance is subject to several types of risks. There are risks that are internal to the project as well as risks that are related to external lenders. Yescombe (2002) stated that projects' financial risks are related to commercial viability, project completion on time and budget (so that lenders start receiving payment), level of anticipated operating revenues and sponsor support for additional funding if required. In most cases, lenders have to examine all the risks borne by the project company to confirm if these risks are reasonable and acceptable to them. External financial risks are attributed to the environment in which the PFI/PPP scheme operates. Other financial risks arise from inflation, interest rates, inadequate hedging of revenue streams and financing costs. In cases where interest rate variation is an issue prior to financial close, NAO (1999a) advises the procuring authorities to take the following action:

- Bidders ought to price their bids on the basis of a reference market interest rate.
- Bidders should provide details on how movements in interest rates are treated.
- Bidders and their lenders should demonstrate that they have sufficient equity capital available to absorb the level of risk transfer being sought.

It is well known that each PFI/PPP project stakeholder has a different perspective on risk. For example, lenders use the following strategies to mitigate risks which might affect the return from their investment (Fight 2006):

- Costs incurred before construction completion are without recourse to lenders for additional funds.
- There is recourse to some project participants for delay and completion costs if the project is abandoned and if the required threshold standards are not achieved.
- Certainty of future revenue streams to service debt.
- The project maximises revenue while minimising operational costs.

Once the project starts operating, the financial risks to the lenders will be significantly reduced from their peak in the contraction and commissioning phases. Other important risks include issues such as refinancing of the project, the stability of the local currency, taxation issues, country/political risks, etc. For further in-depth explanation of these specialised risks, readers can consult Fight (2006) and Yescombe (2002).

Allocation of availability and performance risks

OGC (2002b) defined this risk as:

> The supplier takes on the risk of absorbing the cost and resourcing consequences of making the service available when required, to the agreed levels of performance. Incentive mechanisms can be developed that reward the private sector in return for performance gains.

Availability and performance risks normally include issues such as latent defects, subcontractors' performance, failure to meet standards, default by contractor/ subcontractors, availability of facilities, etc. Availability and maintenance risks are connected. This is because the project company is fully responsible for the maintenance of the asset. Most of these risks are transferred to the private sector. Table 13.1 shows that the client only retains the risk for changes in specification and shares the risks related to relief events, *force majeure* and termination due to *force majeure*. Details on how these issues are dealt with under a typical project agreement are found in Chapter 11. The following principles are normally used in the allocation of this risk:

- Unitary payment must be structured to reflect any variations in service availability and standards.
- Unitary payment is related to both physical availability and performance requirements.
- Availability measurement varies significantly from project to project. Hence, a non-ambiguous measurement definition is essential before risk allocation.
- In rare circumstances (e.g., a *force majeure* event), the private sector may be untitled or relieved from unavailability and lack of performance.

Allocation of variability of revenue risk

Variability of revenue risk is related to both demand and usage risks. Demand risk is attributed to the fact that the demands for services do not match the forecasted levels. The usage risk is due to the fact that the actual usage of the services varies from the level planned for. Variability of revenue risks is related to issues of poor performance, changes in size/volume of provision and demand, unexpected changes in technology, and incorrect estimation of income generation. From a project finance point of view, revenue risk is the risk that returns to debt and equity will be insufficient. Table 13.1 shows which variability of revenue risks are transferred to the private sector and which are retained by the client. As can been seen from the table, the majority of these risks are retained by the client. NAO (1999a) pointed out that it is likely that value for money can only be achieved if most, if not all, of these risks are allocated to the public sector. The circumstances where the private sector is expected to share some of the variability of risks include:

- when the procurer is unable to influence the volume of usage/demand;
- when costs of service provision are based on volume/usage;
- when the design may influence demand;
- when there is little competition;
- when the project is financed by a high proportion of equity investment.

According to the OGC (2002b), usage or volume risks (fluctuations in service use) can be transferred to the private sector by relating the unitary charge to the level of use made of a service. The private sector may require a higher price for managing usage and demand risks. Hence a minimum threshold of usage may have to be guaranteed to the project company so that the revenue risks are reduced to an optimal level.

Allocation of residual value risks

The residual value risk is the risk that the economic value of the asset diverges, either during or at the end of the contract term, from the contracted value. The main concern here is the certainty of the value and liability of physical assets at the end of contract. Normally, this risk is taken by the party taking over the ownership of the asset at the end of the concession period. However, most PFI/PPP projects are highly specialised and it may not be possible to find a secondary market for them or an alternative usage. This may lead to a high liability risk for the party owning the asset. In some circumstances the private sector may require a higher price for bearing this risk. Hence, it probably is not good value for money to transfer residual value risks to the project company. Scenarios on how to deal with a particular facility at the end of the project agreement period are outlined in Chapter 11.

Allocation of termination risks

Termination risks are due to default by the procuring entity as well as default by the operator leading to the financiers stepping in. These scenarios might happen only in narrowly defined circumstances, For example, in cases where the PFI/PPP deal is underpriced by the contractor, termination occurs due to *force majeure*, and events of default. Under the project agreement the project company is normally required to bear costs related to a step-in by lenders; however, clients might face additional costs for management time and advisers' fees. The client may also need to consider other additional costs that may be required to complete service delivery with another contractor. Private contractors are entitled to termination with compensation in situations where sovereign risk occurs. Events where contractors are entitled for reimbursement include:

- when the client is in material breach of his obligations under the project agreement and such a breach shall materially adversely affect the ability of the project company to perform its obligations for a specified period (normally defined by contracts);
- when the client fails to pay any due sum (the sum should not be in dispute) to the project company and such failure continues for a specified period of time;
- when a change in law is made after the effective date;
- when there is political interference and a change of policy and objectives.

The compensation to be paid by the client to the project company is defined by the project agreement schedules. The NAO (1999a) has the following advice for the procuring authorities:

> Departments should evaluate the risks attached to any arrangements which might require them to make such compensation payments . . .
>
> The contracts should be structured to ensure they eliminate all potential upside for the contractor and their lenders if the contract is terminated early through the contract's default or through other action by the contractor other than where the public sector has been in breach of the contract. At all times the contractor and their lenders should be incentivised to continue to provide the services.

Termination risks can be managed effectively through governance, project management and operational monitoring arrangements. These strategies ought to provide PFI/PPP stakeholders with an early warning of any circumstances that could result in a potential termination situation.

Allocation of technology/obsolescence risks

Technology/obsolescence risk is the risk that building assets and equipment will be superseded by technical changes which may result in services being provided using non-effective buildings and technology. All PFI/PPP contracts are over a 30-year period, during which equipment and building components need to be updated. Also investment to acquire new equipment based on the latest technology for the efficient operation of the services, both core and building, should be envisaged under any PFI/PPP deal. If these technical changes result in the purchaser revising their output specification, then they will bear all the associated risk costs. All renovation and renewal of building components and equipment (mechanical services) risks are contracted out to the operating contractor. If the asset is not usable by the end of the concession agreement, the contractor will bear the risk (not receiving a transfer payment) of the asset becoming technologically obsolete. The price of transferring the risks of IT and specialist equipment (such as medical equipment in hospitals) may be beyond the control of the operating company and in most cases this risk is borne by the client.

Limitation of liability

Negligence and duty of care are the fundamental liability principles on which contract law is based. The basic principles of risk allocation are given above on p. 276, but in law there can be risks and liabilities that are not transferable. For example, under safety laws there will be duties of reasonable care on designers, contractors, suppliers, installers, manufacturers, maintainers, operators, and individuals, which will continue unaffected (by the risk transfer under the project agreement) or be shared. It does not matter who takes overall responsibility for the risks associated with aspects of the PFI/PPP projects but this does not free other parties from negligence, duty of care and safety responsibilities. Risk liability in PFI/PPP schemes may arise out of:

- an act or omission which is subject to claim;
- taxation in respect of any termination payment;
- breach of the tenant's covenants;
- liquidation and receivership;
- professional indemnity insurance;
- manufacturer's warranties;
- performance obligation;
- defects;
- technical or professional advice;
- damages payment;
- pollution or contamination;
- ownership and possession;
- third parties.

Under the PFI/PPP agreement, the client has no liability to the building contractor in respect of sums due under the building contract. Also the contractor has no liability to the client under the project agreement for delay in completion of the works. According to Norfolk and Norwich Health Care NHS Trust (2004) the building contractor accepts liability for the following:

- the cost of making good any defects in the works or any part of the works or any physical damage caused to the works as a result of such a defect;
- the reasonable professional fees incurred by the client arising from any such defect in the works or physical damage.

The project company may be relieved from liability to the extent that, by reason of *force majeure*, it is unable to perform its obligations under the project agreement. To avoid lengthy litigation, effective risk control measures are needed and risk allocation should be established at the early stages of the bidding process. It is important that the project's stakeholders understand, agree and act on their risk responsibilities. Further details on liability of the PFI/PPP stakeholders can be found in any standard project agreement.

Dispute resolution mechanisms

Dispute resolution typically involves arbitration in a neutral pursuant to a pre-specified standards and procedures. The Centre for Effective Dispute Resolution (2006) has defined dispute as:

> Any difference or dispute between the Authority and the Provider arising out of or in connection with the Contract (including any question as to the validity or interpretation of the Contract and including any dispute arising before or after termination of the Contract).

According to NAO (2001), the reasons cited for a dispute arising include:

- misinterpretation of the contract;
- poor performance/quality of service/output specification;
- failure to agree prices for new additional services;
- delays/late delivery, missed deadlines;
- changes in requirement;
- disagreement over responsibilities;
- poor communication.

As can be seen from the above dispute list, it seems that the majority of disputes are rooted in the complex contractual provisions that surround the majority of PFI/PPP schemes. In the UK, dispute resolution is addressed by the OGC document 'standard of PFI contract' under clause 27. The procedures and stages for dispute resolution are also listed by NAO (2001) as follows:

- Internal resolution between the authority and contractor. According to OGC guidance, this should be done within a specified period of time (normally seven days). If this fails, then the following action is taken.
- Provision for independent expert resolution or 'Adjudicator'. The 'Adjudicators' are selected from two technical and facilities panels which consist of three senior members of the engineering profession or a similar background. In some cases, financial and clinical services panels are also used for this purpose. If either party disagrees with the expert's decision within 28 days of receipt of the Adjudicator's decision, the dispute is referred to arbitration.
- Provision for the dispute to go to arbitration. The Arbitrator is to deliver his or her decision within 28 days of concluding any hearings which may

have been held in connection with the matter. The decision is final and binding on both parties.

- The case goes to the courts for a final and binding decision. The OGC guidance does not consider this step to be necessary.

The dispute resolution mechanism in the NHS Standard Form Contract for health sector PFI/PPP projects is slightly different from the OGC guidance. The NHS procedure stipulates that the dispute is referred to a liaison committee for resolution first. If the parties have been unable to resolve the dispute within a specified period of time, the parties may refer the dispute to mediation. If this fails, the dispute is referred to the Chief Executives of the Trust and the project company. If this fails, the matter is referred to an expert panel. Finally, if the dispute cannot be resolved by the experts, the matter can be referred to arbitration. The Arbitrator's decision is final and binding.

Other characteristics of dispute resolutions under PFI/PPP procurement systems include the following:

- No reference is made to mediation.
- Disputes during the construction phase are dealt with under the statutory right of adjudication, which applies to building sub-contracts, but not to the project agreement.

In comparing the guidance on dispute resolution provided by the OGC, HNS and the Centre for Effective Dispute Resolution, it appears that the latter is more robust and provides flexibility and equity for all involved parties, which is essential for maintaining good business relationships and partnership.

Summary

The PFI/PPP initiative aims to minimise risk to the public sector and to give the private sector incentives to deliver cost-effective and efficient services. As explained in this chapter, these aims are accomplished by placing risks with the party best able to manage them. This chapter has enabled the reader to understand what the typical risks are in any PFI/PPP construction project.

Bibliography

4Ps (2004a) *Project Information Briefing: Castle Hill Primary School and Joint Service Centre*, London: Public–Private Partnerships Programme.

4Ps (2004b) *PFI Agreement Payment Mechanism*, London: Building Schools for the Future.

4Ps (2006) *Procurement Option: PFI/PPP*, available at: http://www.4ps.co.uk/home.aspx, accessed June 2006.

Accounting Standards Board (1998) *Responses for the Public Record to Exposure Draft: Amendment to FRS 5 'Reporting the Substance of Transactions': The Private Finance Initiative*, London: Accounting Standards Board.

Aggarwal, R. (1980) 'Corporate use of sophisticated capital budgeting techniques: a strategic perspective and a critique of survey results', *Interfaces*, vol. 10, no. 2, pp. 31–34.

Akalu, M. (2001) 'Re-examining project appraisal and control: developing a focus on wealth creation', *International Journal of Project Management*, vol. 19, pp. 375–383.

Akintola, K. *et al.* (1998) 'Risk analysis and management of private finance initiative projects', *Engineering Construction and Architectural Management*, vol. 1, pp. 9–21.

Akintoye, A., Beck, C. and Hardcastle, C. (2003) *Public–Private Partnerships: Managing Risks and Opportunities*. Oxford: Blackwell.

Akintoye, A., Beck, C., Hardcastle, C., Chinyio, E. and Asenova, A. (2001a) 'Risk mitigation practices under PFI environment', in *Proceedings of the RICS Foundation Building Research COBRA Conference*, Glasgow, 3–5 September.

Akintoye, A., Beck, C., Hardcastle, C., Chinyio, E. and Asenova, A. (2001b) *Framework for Risk Assessment and Management of Private Finance Initiative*, Glasgow: Glasgow Caledonian University.

Akintoye, A., Taylor, C. and Fitzgerald, E. (1998) 'Risk analysis and management of private finance initiative projects', *Journal of Engineering, Construction and Architectural Management*, vol. 5, no. 1, pp. 9–21.

Allee, V. (2000) 'The value evolution: addressing larger implications of an intellectual capital and intangibles perspective', *Journal of Intellectual Capital*, vol. 1, no. 1, pp. 17–32.

Alni, M., Petersen, A. and Chapman, K. (2000) 'Applications of a developed quantitative model in building repair and maintenance: case study', *Facilities*, vol. 19, no. 5/6, pp. 215–221.

Architecture: Seminar Series Notes, London: Architectural Association.

Arditi, D. and Gunaydin, H. (1997) 'Total quality management in the construction process', *International Project Management*, vol. 15, no. 4, pp. 235–243.

Arthur Andersen and Enterprise LSE (2000) 'Value for money drivers in the private finance initiative', available at: www.ogc.gov.uk.

Ashworth, A. (2004) *Cost Studies of Buildings*, 4th edn. Harlow: Longman Scientific.

Ashworth, A. and Skitmore, M. (1991) *Accuracy in Estimating*, Occasional Paper No. 27, London: The Chartered Institute of Building.

Atkinson, R. (1999) 'Project management: cost, time and quality, two best guesses and a phenomenon, it's time to accept other success criteria', *International Project Management*, vol. 17, no. 6, pp. 337–342.

The Audit Commission (1988) *Saving Energy in Local Government Buildings*, London: HMSO.

Audit Scotland (2002) *Taking the Initiative: Using PFI Contracts to Renew Council Schools*, http://www.audit-scotland.gov.uk.

Balfour Beatty (2005) *Project Structure*, http://www.balfourbeattyppp.com/balfourbeattyppp/.

Ball, R., Heafey, M. and King, D. (2000) 'Private finance initiative: a good deal for the public purse or a drain on future generations?', *Policy and Politics*, vol. 29, no.1, pp. 95–108.

Bannister, F. (2001) 'Dismantling the silos: extracting new value from IT investments in public administration', *Journal of Information Systems*, vol. 11, pp. 65–84.

Barber, E. (2003) 'Benchmarking the management of projects: a review of current thinking', *International Project Management*, vol. 22, no. 6, pp. 301–307.

Barnes, W. (1983) 'How to allocate risks in construction contracts', *Project Management*, vol. 1, no. 1, p. 24–28.

Bennett, J. and Ferry, D. (1987) 'Towards a simulated model of the total construction process', *Building Cost Modelling and Computers*, London: E & FN Spon, pp. 377–385.

Bent, F., Holm, S. and Søren, M. (2002) 'Underestimating costs in public works projects: error or lie?' *American Planning Association Journal*, vol. 68, no. 3, pp. 279–295.

Betts, M. (1992) 'Financial control of public- and private-sector construction projects in Singapore', *International Project Management*, vol. 10, no. 1, pp. 3–10.

Bing, L. *et al.* (2004) 'The allocation of risk in PPP/PFI construction projects in the UK', *International Journal of Project Management*, vol. 12, no. 1, pp. 17–22.

Birnie, J. (1999) 'Private Finance Initiative (PFI) – UK construction industry response', *Journal of Construction Procurement*, vol. 5, no. 1, pp. 5–14.

Blois, K. (2003) 'Using value equations to analyse exchanges', *Marketing Intelligence and Planning*, vol. 21, no. 1, pp. 16–22.

Bourguignon, A. (2005) 'Management accounting and value creation: the profit and loss of reification', *Critical Perspective on Accounting*, vol. 16, no. 4, pp. 353–389.

Boussabaine A.H. (1996) *Neurofuzzy Modelling of Cost and Duration of Construction Projects*, (EPSRC research grant, GR/K/85001), Research Report No. 4, Liverpool: School of Architecture, University of Liverpool.

Boussabaine, A.H. and Elhag, T. (1999a) *Statistical Analysis and Cost Models Development for Tender Price Estimation*, London: Royal Institution of Chartered Surveyors (EPSRC grant, GR/K/85001).

Boussabaine, A.H. and Elhag, T. (1999b) *Tender Price Estimation Using ANN Methods* (EPSRC research grant, GR/K/85001) Research Report No. 3.

Boussabaine, A.H. and Kirkham, R.J. (2004) *Whole Life Cycle Costing: Risk and Risk Responses*, Oxford: Blackwell Science.

Boussabaine, A.H. and Kirkham, R.J. (2006) 'Whole life cycle performance measurement re-engineering for the UK National Health Service estate', *Facilities*, vol. 24, no. 9/10, pp. 324–342.

Boussofiane, A., Martin, S. and Parker, D. (1997) 'The impact on technical efficiency of the UK privatization programme', *Applied Economics*, vol. 29, pp. 297–310.

Bowman, C. and Ambrosini, V. (2000) 'Value creation versus value capture: towards a coherent definition of value in strategy', *British Journal of Management*, vol. 11, pp. 1–55.

Broadbent, J. and Haslam, C. (2000) 'The origins and operation of the Private Finance Initiative', in *Private Finance Initiative: Saviour, Villain or Irrelevance?* www.ippr.org.uk

Broadbent, J. and Laughlim, R. (2003) 'Public–private partnerships: an introduction', *Accounting, Auditing and Accountability Journal*, vol. 16, no. 3, pp. 332–341.

BSI (2002) *Building and Constructed Assets Service Life Planning*, ISO DIS 1586–5, BSI, UK.

CABE (2002) *The Value of Good Design: How Buildings and Spaces Create Economic and Social Value*, London: Commission for Architecture and the Built Environment.

Capper, P. (2004) 'Why use standard forms of contract?' http://www1.fi' dic.org/resources/contracts/ibc_feb2004/capper_feb04.asp.

Central Unit of Procurement (CUP) (1991) *Specification Writing*, London: Central Unit for Procurement.

Central Unit of Procurement (CUP) (1992) *Life Cycle Costing*, London: Central Unit of Procurement.

Cesare, M.A., Santamarina, C., Turkstra, C. and Vanmarcke, E.H. (1992) 'Modelling bridge deterioration with Markov chains', *Journal of Transportation Engineering, American Society of Civil Engineers*, vol. 118, no. 6, pp. 820–833.

Cha, H. and O'Connor, J. (2005) 'Optimizing implementation of value management processes for capital projects', *ASCE Journal of Construction Engineering and Management*, vol. 131, no. 2.

Chang, H., Cheng, M. and Das, S. (2004) 'Hospital ownership and operating efficiency: evidence from Taiwan', *European Journal of Operational Research*, vol. 159, pp. 513–527.

Chen, A.N., Hwang, Y. and Shao, B. (2005) 'Measurement and sources of overall and input inefficiencies: evidences and implications in hospital services', *European Journal of Operational Research*, vol. 161, no. 2, pp. 447–468.

Chen, Q and Liu, G. (2003) 'Critical success factors for value management studies in construction', Journal of Construction Engineering and Management, vol. 129, no. 5, pp. 485–491.

Choobineh, F. and Behrens, A. (1992) 'Use of intervals and possibility distributions in economic analysis', *Journal of Operational Research Society*, vol. 43, no. 9, pp. 907–918.

City Hospital NHS Trust (2006) *Project Agreement and Schedules*, London.

Clark, G. and Root, A. (1999) 'Infrastructure shortfall in the United Kingdom: the Private Finance Initiative and government policy', *Political Geography*, vol. 18, no. 3, pp. 341–365.

Construction Industry Council (CIC) (2000) *The Role of Cost Savings and Innovations in PFI Projects*, London: Thomas Telford.

Crosby, L.A., Evans, K. and Cowles, D. (1990) 'Relationship quality in services selling: an interpersonal influence perspective', *Journal of Marketing*, vol. 54, no. 3, pp. 68–81.

Dann, N. and Worthing, D. (2005) 'Heritage organisations and conditions surveys', *Structural Survey*, vol. 23, no. 2, pp. 91–100.

de Lemos, T., Betts, M., Eaton, D. and de Almeida, L. (2001) 'Model for management of whole life cycle risk uncertainty in the Private Finance Initiative (PFI)', *Journal of Project Finance*, vol. 7, pp. 1–13.

Department of Health (2004a) *Standard Form Project Agreement*, Version 3, London: DoH.

Department of Health (2004b) *The Design Development Protocol for PFI Schemes, Revision 1*, London: DoH.

Department of Health (2004c) *Public–Private Partnerships in the NHS: The Design Development Protocol for PFI Schemes*, London: DoH.

Design quality indicators by CABE (2005) http://www.dqi.org.uk/reception/library/DQIOnline.PDF, accessed 15.04.05

Dikmen, I., Birgonul, M. and Artuk, S. (2005) 'Integrated framework to investigate value innovations', *Journal of Management in Engineering*, vol. 21, no. 2, 1 April.

Dumond, E.J. (2000) 'Value management: an underlying framework', *International Journal of Operations and Production Management*, vol. 20, no. 9, pp. 1062–1077.

Edwards, P. and Shaoul, J. (2002) 'Partnerships: for better, for worse?', *Accounting, Auditing and Accountability Journal*, vol. 16, no. 3, pp. 397–421.

Edwards, P., Shaoul, J., Stafford, A. and Arblaster, L. (2004) 'Evaluating the operation of PFI in roads and hospitals', *ACCA Research Report* no. 84, Executive Summary, London: Certified Accountants Educational Trust.

Elhag, T. and Boussabaine, A.H. (1997) *Factors Affecting Cost and Duration* (EPSRC GR/K/85001), Research Report No. 1, Liverpool: University of Liverpool.

Elhag, T. and Boussabaine, A.H. (1998) *Statistical Analysis and Cost Models Development for Construction Projects*, (EPSRC GR/K/85001), Research Report No. 2, Liverpool: University of Liverpool.

Esty, B.C. (2002) *An Overview of Project Finance, 2001 Update*, Cambridge, MA: Harvard Business School.

Esty, B.C. (2003a) *Modern Project Finance: A Casebook*, New York: John Wiley & Sons, Inc.

Esty, B.C. (2003b) *The Economic Motivations for Using Project Finance, 2002 Update*, Boston: Harvard Business School.

European International Contractors (2003) *EIC Contractor's Guide to the FIDIC Conditions of Contract for EPC Turnkey Projects ('Silver Book')* 2nd edn, available at: http://www.eicontractors.de//seiten/bookshop/silver_desc.htm.

Farrell, L.M. (2003) 'Principal-agency risk in project finance', *International Project Management*, vol. 21, pp. 547–561.

Ferry, D. and Flanagan, R. (1991) *Life Cycle Costing: A Radical Approach*, London: Construction Industry Research and Information Association.

Fight, A. (2005) *Introduction to Project Finance*, London: Butterworth-Heinemann.

Finlay, D. (2003) *Risk and Reward in PFIs*, London: National Audit Office.

Fitch Ratings (2003) *PPP-PFI: UK Market Trends and Fitch Rating Criteria for European PPP Transactions*, 13 May.

Flanagan, R., Kendell, A., Norman, G. and Robinson, G.D. (1987) 'Life cycle costing and risk management', *Construction Management and Economics*, vol. 5, no. 4, pp. S53–S71.

Flanagan, R. and Norman, G. (1983) *Life Cycle Costing for Construction*, London: Surveyors Publications, RICS.

Flanagan, R. and Tate, B. (1997) *Cost Control in Building Design*, Oxford: Blackwell Science.

Flanagan, R. *et al.* (1989) *Life Cycle Costing: Theory and Practice*, Oxford: BSP Professional.

French, N. and Wiseman, G. (2003) 'The price of space: the convergence of value in use and value exchange', *Journal of Property Investment and Finance*, vol. 21, no. 1, pp. 23–30.

Ford, J. and Shaoul, J. (2001) 'Appraising and evaluating PFI for NHS hospitals', *Financial Accountability and management*, vol. 17, no. 3, pp. 246–270.

Froud, F. (2003) 'The private initiative: risk, uncertainty and the state', *Accounting, Organizations and Society*, vol. 28, no. 6, pp. 567–589.

Fung, J. and Makis, V. (1997) 'An inspection model with generally distributed restoration and repair times', *Micro-electron Reliability*, vol. 37, no. 3, pp. 381–389.

Gadd, S. (2001) *Management of Risk in PFU Projects*, Bristol: Defence Procurement Agency, Private Finance Unit, Issue 2.

Gaffney, D. and Pollock, A. (1999) 'Pump-priming the PFI: why are privately finance schemes being subsidised?' *Public Money Management*, no. January–March, pp. 55–62.

Gaffney, D., Pollock, A.M., Price, D. and Shaoul, J. (1999) 'PFI in the NHS – is there an economic case?', *British Medical Journal*, vol. 319, pp. 116–119.

Gardiner, D. and Stewart, K. (2000) 'Revisiting the golden triangle of cost, time and quality: the role of NPV in project control, success and failure', *International Project Management*, vol. 18, pp. 251–256.

Griffith, A. and Headly, J.D. (1998) 'Management of small building works', *Construction Management and Economics*, vol. 16, pp. 703–709.

Grimsey, D. and Lewis, M.K. (2002) 'Evaluating the risks of public–private partnerships for infrastructure projects', *International Journal of Project Management*, vol. 20, no. 2, pp. 107–118.

Grimshaw, D., Vincent, S. and Willmott, H. (2002) 'Going privately: partnership and outsourcing in UK public services', *Public Administration*, vol. 80, no. 3, pp. 475–502.

Groth, J., Byers, S. and Bogert, J. (1996) 'Capital, economic returns and the creation of value', *Management Decision*, vol. 34, no. 6, pp. 21–30.

Grout, P. (1997) 'The economics of the private finance initiative', *Oxford Review of Economic Policy*, vol. 13, no. 4, pp. 53–66.

Guzhva, V.S. and Pagiavlas, N. (2003) 'Corporate capital structure in turbulent times: a case study of the US airline industry', *Journal of Air Transport Management*, vol. 9, pp. 371–379.

Hall, M., Holt, R. and Purchase, D. (2003) 'Project sponsors under New Public Management: lessons from the frontline', *International Project Management*, vol. 21, no. 56, pp. 495–502.

Harris, F. and McCaffer, R. (1989) *Modern Construction Management*, Oxford: BSP Professional Books.

Hawksworth, J. (2000) 'Implications of the public sector financial control framework for PPPs', in *Private Finance Initiative: Saviour, Villain or Irrelevance?* London: Commission on Public–Private Partnerships.

Higher Education Funding Council for England (HEFCE) (1998) *Practical Guide to PFI for Higher Education Institutions*, London: Higher Education Funding Council for England.

HM Treasury (1993) *Private Finance Initiative: Breaking New Ground*, London: HM Stationery Office.

HM Treasury (1994) *Government Procurement: Progress Report to Prime Minister 1993–94*, London: HM Stationery Office.

HM Treasury (1997a) *Partnerships for Prosperity*, London: HM Stationery Office.

HM Treasury (1997b) *The Green Book: Appraisal and Evaluation in Central Government*, London: HM Stationery Office.

HM Treasury (1998) *Constructing the Best Government Client: Pilot Benchmarking Study*, London: HM Stationery Office.

HM Treasury (1999a) *How to Construct a Public Sector Comparator, Series 3*, Technical Notes, London: Private Finance Treasury Taskforce, HM Treasury.

HM Treasury (1999b) *How to Account for PFI Transactions*, Technical Note no. 1, London: Private Finance Treasury Taskforce, HM Treasury,

HM Treasury (2000a) *Public–Private Partnerships: The Government's Approach*, London: The Stationery Office.

HM Treasury (2000b) *Value for Money Drivers in the Private Finance Initiative*. (a report by Arthur Andersen and Enterprise LSE) London: Private Finance Treasury Taskforce, HM Treasury.

HM Treasury (2003a) *PFI: Meeting the Investment Challenge*, London: HM Treasury.

HM Treasury (2003b) *The Green Book: Appraisal and Evaluation in Central Government*, London: HM Treasury.

HM Treasury (2003c) *Supplementary Green Book: Guidance on the Treatment of Optimism Bias*, London: HM Treasury.

HM Treasury (2004a) *Standardisation of PFI Contracts*, Version 3, London: The Stationery Office.

HM Treasury (2004b) *The Green Book: Appraisal and Evaluation in Central Government*, rev. edn, London: HM Treasury.

HM Treasury (2005) *Application Note: Value for Money in Refinancing*, London: HM Treasury.

Holmes, R. (1994) *CIOB Handbook of Facilities Management*, ed. Alan Spedding, Harlow: Longman.

Hood, J. and McGarvey, N. (2002) 'Managing the risks of public–private partnerships in Scottish local government', *Policy Studies*, vol. 23, no. 1, pp. 21–35.

House of Commons Select Committee on Treasury (2002) *Fourth Report on the Private Finance Initiative*, http://ww.publications.parliament.uk/pa/cm199900/cmselect/cmtreasy147/14703.htm, accessed June 2006.

Humphreys, K. (1991) *Cost and Optimization Engineering*, 3rd edn, New York: McGraw-Hill.

Jackson, P. (2001) 'Public sector added value: can bureaucracy deliver?' *Public Administration*, vol. 79, no. 1, pp 5–28.

Johnson, J. (2005) 'Can complexity help us to better understand risk?' in A.H.

Boussabaine, W. Lewis, R.J. Kirkham and? Jared (eds) *Proceedings of the 1st International Conference on Built Environment Complexity*, Liverpool.
Kelly, J. and Male, S. (2002) 'Value management', in J. Kelly, R. Morledge and S. Wilkinson (eds) *Best Value in Construction*, Oxford: Blackwell, pp. 77–99.
Kent, D. (2003) 'Public–private partnerships in prison construction and management', in B. Dinesen and J. Thompson (eds) *Overview of the MPA's 21st Annual Conference*, Major Projects Association, http://www.majorprojects.org/pubdoc/690.pdf.
Kirk, S.J. and Dell'Isola, A.J. (1996) *Life Cycle Costing for Design Professionals*, New York: McGraw-Hill.
Kirkham, R.J., Boussabaine, A.H. and Awwad, A.H. (2002) 'Probability distributions of facilities management costs in NHS acute care hospital buildings', *Construction Management and Economics*, vol. 20, no. 3, pp. 251–261.
KPMG (2005) *Report on Effectiveness of Operational Contracts in PFI*, Report number 215–054, London: KPMG.
Langston, C. (2005) *Life Cost-Approach to Building Evaluation*, NSW: University of New South Wales Press.
Lapierre, J. (1997) 'What does value mean in business-to-business professional services?' *International Journal of Service Industry Management*, vol. 8, no. 5, pp. 377–397.
Lefley, F. (1997) 'Approaches to risk and uncertainty in the appraisal of new technology capital projects', *International Journal of Production Economics*, vol. 53, pp. 21–33.
Lefley, F. and Sarkis, J. (1997) 'Short-termism and the appraisal of AMT capital projects in the US and UK', *International Journal of Production Research*, vol. 35, no. 2, pp. 341–368.
Leiasshvily, P. (1996) 'Towards the teleological understanding of economic value', *International Journal of Social Economics*, vol. 23, no. 9, pp. 4–14.
Leifer, D. (1997) 'A basis for the economic appraisal of intelligent building proposals', *Proceedings of IBC/IC Intelligent Building Congress*, pp. 9–17.
Lewendon, R. (2006) 'The use of NEC: engineering and construction contract', http://www.waterways.org.uk/library/restoration/tech_handbook/Chap%2021.pd, accessed May 2006.
Li, H., Cheng, E., Love, P. and Irani, Z. (2001) 'Co-operative benchmarking: a tool for partnering excellence in construction', *International Project Management*, vol. 19, no. 7, pp. 171–179.
Li, H., Shen, Q. and Love, P. (2005) 'Cost modelling of office buildings in Hong Kong: an exploratory study', *Facilities*, vol. 23, no. 9/10, pp. 438–452.
Lovata, L. and Costigan, M. (2002) 'Empirical analysis of adopters of economic value added', *Management Accounting Research*, vol. 13, pp. 215–228.
MacDonald and Mott (2002) *Review of Large Public Procurement in the UK: A Report on Optimism in Public Procurement*, commissioned by HM Treasury, UK.
Major Projects Association (2003) 'PFI/PPP projects: are they working?' in B. Dinesen and J. Thompson (eds) *Overview of the MPA's Annual Conference*, Major Projects Association.http://www.majorprojects.org/pubdoc/690.pdf.
Malmi, T. and Ikäheimo, S. (2003) 'Value based management practices: some evidence from the field', *Management Accounting Research*, vol. 14, no. 3, pp. 235–254.
Manual (2001) *Project Finance: Introductory Manual on Project Finance for Managers of PPP Projects*, Johannesburg: National Treasury of South Africa.
Maunder, D, and Haggard, M. (2003) 'Delivering the solutions? Financing "new

technologies" ', paper presented at Waste 2004 Conference by Impax Capital and Entec, Stratford-upon-Avon.

McDowall, E. (1999a) 'Linking PFI performance to payment', *Facilities Management*, October, pp. 8–9.

McDowall, E. (1999b) 'Specifying performance for PFI, *Facilities Management*, June, pp. 10–11

McDowall, E. (2000) 'Selecting a preferred bidder, *Facilities Management*, July, pp. 8–9.

McGeorge, D. and Palmer, A. (2002) *Construction Management: New Directions*, Oxford: Blackwell.

Miller, R. and Lewis, W. (1991) 'A stakeholder approach to marketing management using the value exchange models', *European Journal of Marketing*, vol. 25, no. 8, pp. 55–68.

Mills, E.D. (1994) *Building Maintenance and Preservation: A Guide for Design and Management*, 2nd edn, Oxford: Butterworth-Heinemann.

Modanoglu, M., Chand, D. and Chu, Y. (2004) 'Creating economic value in the US airline industry: are we missing the flight?' *International Journal of Contemporary Hospitality Management*, vol. 16, no. 5, pp. 294–298.

Morley, A. (2002) 'The economic benefits of infrastructure projects procured with private finance', paper presented at International Congress, Washington, DC, 19–26 April.

Nanayakkara, R. (2002) *Procurement of Building Services Operation and Maintenance*, DTI Contract no. 38/6/189, London: BSRIA, UK.

NAO (1998) *The Private Finance Initiative: The Four First Design, Build, Finance and Operate Road Contracts*, London: The Stationery Office.

NAO (1999a) *Examining the Value for Money Deals Under the Private Finance Initiative*, London: The Stationery Office.

NAO (1999b) *The PFI Contact for the New Dartford and Gravesham Hospital*, London: The Stationery Office.

NAO (2001) *Managing the Relationship to Secure a Successful Partnership in PFI Projects*, London: The Stationery Office.

NAO (2002) *PFI Refinancing Update*, National Audit Office, HC 1288 2001–2002, London: The Stationery Office.

NAO (2003a) *PFI: Construction Performance*, London: The Stationery Office.

NAO (2003b) *Private Finance Initiative: Redevelopment of MOD Main Building*, London: The Stationery Office.

NAO (2003c) *PFI in Schools*, London: The Stationery Office.

NAO (2005a) *Darent Valley Hospital: The PFI Contract in Action*, National Audit Office, HC 2092004–2005, London: The Stationery Office.

NAO (2005b) *Improving Public Services through Better Construction*, National Audit Office, HC 364 2004–2005, London: The Stationery Office.

Neap, H.S. and Celik, T. (1999) 'Value of a product: a definition', *International Journal of Value-Based Management*, vol. 12, pp. 181–191.

Newton, S. (1991) *An Agenda for Cost Modelling Research, Construction Management and Economics*, London: E & N Spon.

NHS (1994) *Capital Investment Manual*, Leeds: NHS Executive.

NHS (1995) *Private Finance and Capital Investment Projects*, HSG(95)15, Leeds: NHS Executive.

NHS (1999) *Public–Private Partnerships in the National Health Service: The Private Finance Initiative*, Leeds: NHS Executive.

NHS Estates (1994) *Capital Investment Manual*, National Health Service Estates, London. http://www.dh.gov.uk/PolicyAndGuidance/OrganisationPolicy/EstatesAndFacilitiesManagement/EstatesAndFacilitiesArticle.

NHS Estates (2001a) *Estate Code: Essential Guidance on Estates and Facilities Management*, London: National Health Service Estates.

NHS Estates (2001b) *Sustainable Development in the NHS*, London: Department of Health.

NHS Estates (2002) *The Best Client Guide: Good Practice Briefing and Design Manual*, London: Department of Health.

NHS Estates (2004) *The Design Development Protocol for PFI Schemes*, Revision 1, London: Department of Health.

NHS Executive (2000) *The PFI Procurement Process*, London: National Health Service Executive.

NHS Executive (2002a) *Improving PFI Procurement*, London: National Health Service Executive.

NHS Executive (2002b) *Department of Health Standard for Payment Mechanism Schedule 18*, London: National Health Service Executive.

NHS Executive (2003) *Standard Output Specification (Version 2)*, London: National Health Service Executive.

Nijkamp, P. and Ubbels, B. (1999) 'How reliable are estimates of infrastructure costs? A comparative analysis', *International Journal of Transport Economics*, vol. 26, no. 1, pp. 23–53.

Noortwijk, J. and Frangopol, D. (2004) 'Two probabilistic life-cycle maintenance models for deteriorating civil infrastructures', *Probabilistic Engineering Mechanics*, vol. 19, no. 4, pp. 345–359.

Norfolk and Norwich Health Care NHS Trust (2004) *Project Agreement*, amended version, Norwich.

Norfolk and Norwich Hospital (2006) *Facilities Management Agreement*, hospital website, accessed March 2006.

Northcott, D. and Llewellyn, S. (2003) 'The "ladder of success" in healthcare: the UK national reference costing index', *Management Accounting Research*, vol. 14. pp. 51–66.

ODMP (2005) *Best Value in Local Authority Procurement: Guidance Notes*, the website of the Office of the Deputy Prime Minister, available at: www.odpm.gov.uk.

OGC (2001) *PFI List of Signed Projects (as at 30th of September 2001)*, London: Office of Government Commerce.

OGC (2002a) *Principles of Contract Management: Service Delivery*, London: The Stationery Office.

OGC (2002b) 'Best practice: risk allocation in long-term contracts', available at: www.ogc.gov.uk/sdtoolkit/reference/ogc_library/,bpbriefings/risk_allocation.pdf.

OGC (2002c) *Green Public–Private Partnerships*, Norwich: Office of the Deputy Prime Minister.

OGC (2003a) *Achieving Excellence in Construction, Procurement Guide No. 4: Risk and Value Management*, London: Office of Government Commerce.

OGC (2003b) *Whole Life Costing and Cost Management, Procurement Guide no. 7*, London: Office of Government Commerce.

OGC (2004a) *The Gateway Process: Gateway to Success*, London: Office of Government Commerce.

OGC (2004b) *Value for Money Measurement*, OGC Business Guidance, London: Office for Government and Commerce.

OGC (2004c) *The OGC Best Practice Guide: Gateway Reviews*, London: The Office for Government and Commerce and HMSO.

OGC (2005) Successful Delivery Toolkit™ 2005, Version 5.02.

Otley, D. (1999) 'Performance management: a framework for management control systems research', *Management Accounting Research*, vol. 10, no. 363–382.

Palmer, A., Kelly, J. and Male, S. (1996) 'Holistic approach to value engineering in construction in United States', *Journal of Construction Engineering and Management*, vol. 122, no. 4., pp. 324–328.

Patterson, F. and Neailey, K. (2002) 'A risk register database systems to aid the management of project risk', *International Journal of Project Management*, vol. 20, no. 2, pp. 365–374.

Paulin, M., Ferguson, R. and Payaud, M. (2000) 'Effectiveness of rational and transactional cultures in commercial banking: putting client-value into the competing values model', *International Journal of Bank Marketing*, vol. 18, no. 7, pp. 328–337.

Pearce, D. (2003) 'The social and economic value of construction: the construction industry's contribution to sustainable development', www.ncrisp.org.

Peppard, B., Lambert, R. and Edwards, C. (2000) 'Whose job is it anyway?: Organizational information competencies for value creation', *Information Systems Journal*, vol. 10, pp. 291–322.

Picken, H. and Mak, S. (2001) 'Risk analysis in cost planning and its effect on efficiency in capital cost budgeting', *Logistics Information Management*, vol. 14, no. 5/6, pp. 318–327.

Pollock, A., Shaoul, A., Rowland, D. and Player, S. (2001) *Public Services and the Private Sector*, London: Catalyst.

Pollock, A., Shaoul, J. and Vickers, N. (2002) 'Private finance and "value for money" in NHS hospitals: a policy in search of a rationale?' *British Medical Journal* vol. 324, no. 7347, pp. 1205–1209.

PricewaterhouseCoopers (2001) *Public–Private Partnerships: A Clearer View*, London: PricewaterhouseCoopers.

PricewaterhouseCoopers (2002) *Study into Rates of Return Bid on PFI Projects*, London: OGC.

PricewaterhouseCoopers (2003) *Public–Private Partnerships: UK Expertise for International Markets*, London: International Financial Services.

Pringle, J. and Cole, J. (2004) 'The exemplar model: lower costs, better outcomes, in public sector procurement and the public interest', in Chevin, D., *Public Sector Procurement and the Public Interest*, London: The Smith Institute, pp. 89–96.

Private Finance Panel Executive (1992) *Private Finance: Guidance for* Departments, London: HMSO.

Private Finance Panel Executive (1995) *Private Opportunity, Public Benefit: Progressing the Private Finance Initiative*, London: HMSO.

Private Finance Panel Executive (1996a) *Writing an Output Specification*, London: HMSO.

Private Finance Panel Executive (1996b) *Risks and Rewards in PFI Contracts*, London: HMSO.

Private Finance Panel Executive (1996c) *Transferability of Equity*, London: HMSO.

Puig-Junoy, J. (2000) 'Partitioning input cost efficiency into its allocative and technical components: an empirical DEA application to hospitals', *Socio-Economic Planning Sciences Journal*, vol. 34, pp. 199–218.

Raftey, J. (1991) *Models for Construction Cost and Price Forecasting: Proceedings of the First National Research Conference*, London: E & F N Spon.

Raftey, J. (1994) *Risk Analysis in Project Management*, London: E & F N Spon.

RAMP (2006) 'Risk analysis and management for projects', available at: http://www.ramprisk.com/riskmanagement, accessed May 2006.

RIBA (2005) 'RIBA consultation: introducing smart PFI', available at: www.architecture.com, accessed 30 March 2005.

Robinson, P. (2000) 'PFI and the public finances', in *Private Finance Initiative: Saviour, Villain or Irrelevance*? London: Commission on Public–Private Partnerships.

Royal Academy of Engineering (1999) *The Long-Term Costs of Owning and Using Buildings*, London: The Royal Academy of Engineering.

Sandahl, G. and Sjogren, S. (2003) 'Capital budgeting methods among Sweden's largest groups of companies: the state of the art and a comparison with earlier studies', *International Journal of Production Economics*, vol. 84, pp. 51–69.

Sayce, S. and Connellan, O. (2002) 'From existing use to value in use: time for a paradigm shift?', *Property Management*, vol. 20, no. 4, pp. 228–251.

Scheinkestel, N.L. (1997) 'The debt-equity conflict: where does project financing fit?' *Journal of Banking and Finance Law and Practice*, vol. 8, pp. 103–124.

Scherer, W.T. and Glagola, D.M. (1994) 'Markovian models for bridge maintenance management', *Journal of Transportation Engineering, American Society of Civil Engineers*, vol. 120, no. 1, pp. 37–51.

Shaoul, J. (2005) 'A critical financial analysis of the Private Finance Initiative: selecting a financing method or allocating economic wealth?' *Critical Perspectives on Accounting*, vol. 16, no. 4, pp. 441–471.

Shen, Q. and Liu, G. (2003) 'Critical Success Factors for Value Management', *Journal of Construction Engineering and Management*, vol. 129, no. 5, pp. 485–491.

Shillito, M.L. and De Marle, D.J. (1992) *Value: Its Measurement, Design, and Management*, New York: John Wiley & Sons Inc.

Simister, S.J. and Green, S.D. (1997) 'Recurring themes in value management practice', *Engineering, Construction and Architectural Management*, vol. 4, no. 2, pp. 113–125.

Skitmore, M. and Patchell, B. (1990) 'Developments in contract price forecasting and bidding techniques', in P.S. Brandon (ed.) *Q.S. Techniques: New Directions*, Oxford: BPS Professional Books, pp. 75–120.

Smith, J. and Love, P. (2004) 'Auditing construction costs during building design: a case study of cost planning in action', *Managerial Auditing Journal*, vol. 19, no. 2, pp. 259–271.

Snoj, B., Korda, A. and Murnel, D. (2004) 'The relationships among perceived quality, perceived risk and perceived product value', *Journal of Product and Brand Management*, vol. 13, no. 3, pp. 156–167.

Sohail, M., Miles, D. and Cotton, A. (2002) 'Developing monitoring indicators for urban micro contracts in South Asia', *International Project Management*, vol. 20, no. 7, pp. 583–591.

Spackman, M. (2002) 'Public–private partnerships: lessons from the British approach', *Economic Systems*, vol. 26, pp. 283–301.

Stewart, P. (2005) *Sustaining the Benefits of Change in Public Sector Procurement and the Public Sector*, London: The Smith Institute.

Sussex, J. (2001) *The Economics of the Private NHS*, London: Office of Health Economics.

Sussex, J. (2002) *Making the Best of the Private Finance Initiative in the NHS*, London: Office of Health Economics,

Sussex, J. (2003) 'Public–private partnerships in hospital development: lessons from the UK's "Private Finance Initiative" ', *Research in Healthcare and Financial Management*, vol. 2, no. 1, pp. 59–76.

Tanega, J. and Sharma, P. (2000) *International Project Finance: A Legal and Financial Guide to Bankable Proposals*, London: Butterworths.

The Centre for Effective Dispute Resolution (2006) 'Dispute resolution procedure for PFI and long-term contracts', http://www.cedr.co.uk/library/documents/drpforpfi.pdf, accessed May 2006.

The Smith Institute (2005) *Public Sector Procurement and the Public Interest*, ed. D. Chevin, London: The Smith Institute.

The State of Victoria (2001) *Partnerships Victoria Guidance: Risk Allocation and Contractual Issues Guide*. Department of Treasury and Finance, The State of Victoria, Australia.

Then, D. (1995) 'Computer building condition survey', *Facilities*, vol. 13, no. 7, pp. 23–27.

Thomson, D., Austin, S., Wright, H. and Mills, G. (2003) 'Managing value and quality in design', *Building Research and Information*, vol. 31, no. 5, pp. 334–345.

Treasury Taskforce (1999a) *A Step-by-Step Guide to the PFI Procurement Process*, London: HM Treasury.

Treasury Taskforce (1999b) *News No. 177/999*, 29 October.

Treasury Taskforce (1999c) *How to Appoint and Work with a Preferred Bidder*, London: HM Treasury.

Treasury Taskforce (1999d) *How to Achieve Design Quality in PFI Projects*, London: HM Treasury.

Treasury Taskforce (1999e) *How to Account for PFI Transactions*, London: HM Treasury.

Treasury Taskforce (2000a) *How to Manage the Delivery of Long-Term PFI Contracts*, London: HM Treasury.

Treharne, T. (2003) 'Financing and PFI', at http://www.majorprojects.org/pubdoc/690.pdf, accessed 2005.

Tsai, P. and Molinero, C. (2002) 'A variable returns to scale data envelopment analysis model for the joint determination of efficiencies with an example of the UK health service', *European Journal of Operational Research*, vol. 141, pp. 21–38.

Turner, R. and Keegan, A. (1999) 'The versatile project-based organization: governance and operational control', *European Management Journal*, vol. 17, no. 3, pp. 296–309.

Turner, R. and Keegan, A. (2001) 'Mechanisms of governance in the project-based organization: roles of the broker and steward', *European Management Journal*, vol. 19, no. 3, pp. 254–267.

UNISON (2004) *Public Risk for Private Gain? The Public Audit Implications of Risk Transfer and Private Finance*, ed. A. Pollock, D. Price and S. Player, London: UNISON.

Van Horne, J. (1995) *Financial Management and Policy*, 10th edn, London: Prentice Hall International, UK.

Weller, D.L. (1996) 'Benchmarking: a paradigm for change to quality education', *The TQM Magazine*, vol. 8, no. 6, pp. 24–29.

Williams, G.B. and Hirani, R.S. (1997) 'A delay time multi-level on-condition preventative maintenance inspection model based on constant base interval risk – when inspection detects pending failure', *International Journal of Machine Tools Manufacturing*, vol. 37, no. 6, pp. 823–836.

Williams, T. (1994) 'Using a risk register to integrate risk management in project definition', *International Journal of Project Management*, vol. 12, no. 1, pp, 17–22.

Wirahadikusumah, R., Abraham, D.M. and Castello, J. (1999) 'Markov decision process for sewer rehabilitation', *Engineering, Construction and Architectural Management*, vol. 6, no. 4, pp. 358–370.

Woodruff, R.B. (1997) 'Customer value: the next source for competitive advantage', *Journal of the Academy of Marketing Science*, vol. 25, no. 2, pp. 139–153.

Ye, S. and Tiong, R. (2003) 'Tariff adjustment frameworks for privately financed infrastructure projects', *Construction Management and Economics*, vol. 21, pp. 409–419.

Yescombe, E. (2002) *Principles of Project Finance*, Oxford: Elsevier.

Zaring, O. (1996) 'Capital budgeting for the unexpected', *Scandinavian Journal of Management*, vol. 12, no. 3, pp. 233–241.

Index